FLEET FIRE

FLEET FIRE

Thomas Edison and the Pioneers of the Electric Revolution

L. J. DAVIS

ARCADE PUBLISHING
New York

For Bill and Susan

FIRST EDITION

Library of Congress Cataloging-in-Publication Data

  Davis, L. J. (Lawrence J.)
      Fleet fire : Thomas Edison and the pioneers of the electric revolution / L. J.
      Davis. —1st ed.
        p. cm.
      Includes bibliographical references and index.
      ISBN 1-55970-655-4
        1. Electric engineering—History. 2. Electric engineers. I. Title.
  TK15 .D39 2003
  621.3'09—dc21   2002028367

Published in the United States by Arcade Publishing, Inc., New York
Distributed by AOL Time Warner Book Group

Visit our Web site at www.arcadepub.com

10 9 8 7 6 5 4 3 2 1

Designed by API

EB

PRINTED IN THE UNITED STATES OF AMERICA

CONTENTS

ACKNOWLEDGMENTS

I can make certain claims to scholarship, but only a few in the direction of originality. The most prominent of the latter was finding the Rutgers University's abstracts of the notebooks that Edison, Batchelor, and Upton kept during the creation of the light bulb, which clarified many matters and stripped away the stuff of legend in which Edison enwrapped that seminal event. Another was working out just what Tesla thought he was up to during the years 1899–1901, especially because of all the erroneous and misguided things that have been written — and are still being written — about that brilliant, seriously strange man.

First, I owe a debt to my agent, Gordon Kato, who was deeply committed to the project. It was Gordon who fielded a letter from one editor who wrote that she thought Benjamin Franklin invented electricity and Thomas Edison perfected it, and what are all these other people doing in here? I also appreciate the incisive, always helpful editorial queries and comments of both Dick Seaver and Darcy Falkenhagen of Arcade, who labored long in forcing me to rethink and clarify various aspects of this complex and elusive subject.

My brother and sister-in-law, to whom the book is dedicated, were pillars of support. And, of course, I could never have written *Fleet Fire* without the constant help of my lifelong companion, Judith Rascoe.

FLEET FIRE

INTRODUCTION

WHAT WAS KNOWN

The ancients knew that amber, when rubbed, would attract lint, hair, and chaff — the fibrous residue that remained after wheat was threshed — although how long they knew it probably can't be determined. The Greek word for amber was *electron*, for its golden color, although the term was also used for gold and for alloys of silver and gold such as electrum, likewise because of their color. Thales of Miletus, one of the Seven Wise Men of ancient Greece and a founder of the science of mathematics, believed a magnet attracted iron because it had a soul. Sometime around the year 585 B.C., he made the first recorded reference to a lodestone, iron charged with magnetism in its natural state, and its amberlike properties. Chinese literature on the subject begins to appear in the third century B.C., and in ancient Rome both Plinius Secundus, known to history as Pliny the Elder, and Claudius Galen, the great Greek anatomist, addressed the subject. But Galen frankly confessed that no explanation of the phenomenon made any sense to him. Saint Augustine, too, was puzzled by the fact that magnetism attracted iron chains. No one else seems to have given much thought to the mysterious matter and, to a certain extent, the phenomenon remains mysterious to this day. Thales, Pliny, and Galen knew that rubbed amber and an unrubbed lodestone attracted certain seemingly unrelated substances, and that they stuck to certain metals. So do we.

We also know, because we have since discovered it, that magnetism is one of the four basic forces of the universe. The others are gravity, weak force, and electricity. No one knows what gravity is.

Like all basic forces, it appears to be ubiquitous in the universe, and like all forces, it acts visibly on its surroundings. Gravity, for example, keeps people from flying off the earth and the earth from flying away from the sun. Weak force has been theorized, perhaps as having something to do with the dark matter that supposedly accounts for most of the universe's mass, but has never been observed. Neither has dark matter. As far as we know, electricity and magnetism are the only two of the four forces that are convertible into each other. We also know that electricity is a stream of excited electrons, a thing that was not discovered until after 1897, when the electron was identified and described. But on the vexed subject of why magnetism and the other forces occur or where they came from, we were pretty much in the same boat as Thales until William Gilbert, an English medical doctor who became the personal physician of Elizabeth I, took up the subject in 1600, when he published the book *De Magnete*.

Gilbert coined the words *magnetism* and *electricity*, and he was the first to distinguish between the two. This was a crucial insight. Few people even suspected that electricity existed, but those who knew had observed that it was similar to magnetism. That is, like magnetism it seemed to manifest itself (but as sparks and pain, unlike magnetism, which is a passive force) in the presence of metal, hair, and chaff, and like magnetism it had the power to attract and repel. And like the magnetism induced in amber, electricity manifested itself when certain things were rubbed. Scuffling the leather soles of shoes or sandals over a woolen rug produced a spark and pain when the wearer of the shoes or sandals wet a finger and touched a stone wall or the flesh of another person. Defining electricity as a separate phenomenon meant that electricity could be studied as its own entity, and by studying it separately, its uses (if any) could be learned. With the science of 1600, Gilbert could go no further into the nature of electricity than this. We, of course, have gone much further, making electricity one of the underpinnings of our civilization. But like Gilbert, we still have no idea why electricity exists.

Decades before Francis Bacon became the father of the scientific method in the England of the Stuarts, the Tudor-era Gilbert was a firm believer in hands-on experimentation rather than windy speculation based on ancient texts. "Many modern authors," he wrote,

> have written about amber and jet attracting chaff and other facts unknown to the generality: with the results of their labors booksellers' shops are crammed full. Our generation has produced many volumes about recondite, abstruse and occult causes and wonders, and in all of them amber and jet are represented as attracting chaff; but never a proof from experiment, never demonstration do you find in them. The writers deal only in words that involve in thicker darkness subject-matter; they treat the subject esoterically, miracle-mongeringly, abstrusely, reconditely, mystically. Hence such philosophy bears no fruit; for it rests simply on a few Greek or unusual terms — just as our barbers toss off a few Latin words in the hearing of an ignorant rabble in token of their learning, and thus win their reputation.[1]

Here speaks a genuinely modern voice. Gilbert rubbed objects and compiled a huge list of things that could thus be electrified. He employed a sensitive electricity detector that resembled an unmagnetized compass, and he used a true compass to study magnetism, because a true compass always points to the magnetic north pole. He deduced that electrical attraction was different from magnetic attraction — because, ignoring lightning, there appeared to be no naturally occurring forms of electricity in the way that magnetism appeared in a lodestone, a naturally occurring magnet — and he correctly concluded that "the whole earth is a magnet." To explain his discovery to the queen, he built a terrella, or "little earth," a tiny model of the planet made of lodestone, and passed a compass over it. Galileo, who did not suffer fools gladly, was a great admirer of Gilbert's work.

Once Gilbert had demonstrated the uses of the scientific method in the study of electricity, interest in the subject picked up, but not by much. Otto von Guericke, a German lawyer, mathematician, engineer, and burgomaster of Magdeburg for thirty-five years, invented the vacuum pump and built a Gilbert-like terrella of his own, and with further experimentation produced a friction-based electrical generator, the first of its kind. Leather pads were placed in contact with a flat wheel made of glass or sulfur and the wheel was rotated, either with pedals or with a crank. Electricity resulted, if weakly. In 1729 a London pensioner named Stephen Gray managed to transmit electricity along 765 feet of thread, and he may have been the first to use copper as a conductor — that is, a substance that would convey electricity from one place to another. There, however, matters more or less rested until 1746, when a professor of physics in Leyden, Holland, named Pieter van Musschenbroek stored electricity in a foil-wrapped jar of water. His device became known as the Leyden jar, although some scholars claim that van Musschenbroek shares the honor of its invention with his friend Andreas Cunaeus and with the German physicist Ewald Jürgen von Kleist. It worked in this way: A glass jar was filled with water and corked. Through a hole in the cork ran a wire. To the wire was attached a von Guericke friction generator, which was then activated, charging the water with electricity, which could then be used for experimental purposes, such as identifying the substances that would convey electricity. The Leyden jar was the world's first battery, although today we would call it a capacitor, a device that will store electricity but only for a short while, the way a leaky jug stores water. There was a small amount of electricity in there, at least for a while, and you could get to it. No one knew how the trick was done.

Van Musschenbroek proved to himself that he had succeeded in storing electricity when he touched the wire and the outside of the jar at the same time and produced a shock that convinced him "it was all up with me." For years thereafter, French kings amused themselves by using the Leyden jar, improved with layers of metal foil lining the inside and outside of the bottle, to shock long lines of hand-holding clergymen, courtiers, and guardsmen, all of whom gratifyingly leapt into the air at the same time.

As the eighteenth century dawned, the last phase of the ten-thousand-year-old Neolithic Revolution was entering its final phase, at least until fertilizers produced by the new science of chemistry and the agricultural machinery built by the Industrial Revolution improved the landscape again. The Neolithic Revolution had begun with the invention of agriculture, and improvement was, of course, in the eye of the beholder, which was identical with the eye of the man whose ox was gored. Agricultural laborers, whose livelihoods depended largely on work during haymaking and harvest, destroyed many of the new threshing machines. It was a world that had moved slowly for time out of mind, and the lot of most people was not a happy one. The Neolithic Revolution had given them cities, literacy, religion, the printing press, sailing vessels, gunpowder, and a wealth of diseases unknown to folk who hunt and gather. They were clothed, after a fashion, and the home weaving of cloth was England's largest industry. England, indeed, had become the world's largest trading nation, and its agriculture was the most efficient in the world, yielding four to nine bushels of wheat from an acre. The English peasantry was largely a thing of the past, and the yeoman farmer who had formed such an important part of Cromwell's support in the Puritan Revolution was gone as well, although the yeomanry continued to thrive in the American colonies. In America the life expectancy was thirty-five years, in England it was twenty-five, and most of the population of London was drunk most of the time.

In this era there were a few specialized places, such as Coalbrookdale in Shropshire, where an abundance of iron and coal, at least by the era's standards, made possible a concentration of ironworking, but most regions contained the entire infrastructure of civilization, from the production of foodstuffs to the production of clothing. The same was true in the American colonies, except that certain manufactures, such as glass, were banned for the benefit of the manufacturers in the mother country. But unknown to almost everyone, in Britain and America there were Richard Arkwright, James Watt, and Benjamin Franklin, who would begin to set forward processes that would utterly — and, in historical terms, in the blink of an eye — transform their countries and their world. In 1752 the American Franklin invented the lightning rod. In 1769 the Englishman Arkwright assembled the world's first automated spinning mill for the making of cloth, and the Scotsman Watt repaired a pump and in so doing invented the modern steam engine. The world would never be the same. To be sure, in both England and the colonies, most people worked when they had to or when they wanted to, and the minuscule upper classes (which didn't work at all, except in the law courts, Parliament and the colonial legislatures, the bureaucracy, and the universities) believed the countryside was sunk in idleness and sloth. There was some indication that this was so. For example, there was a widespread practice called "Saint Monday," where Monday as well as Saturday and Sunday was a holiday to enable the men to recover from their hangovers or keep on drinking and the women to put their homes and hovels to rights. In some places, there was also a Saint Tuesday.

The economy expanded at a rate somewhere between three-tenths and four-tenths of a percentage point a year, roughly as much as a modern economy expands in a good month. And, in a country with miserable roads, where foodstuffs or anything else could not move around easily, if at all, local conditions dominated the lives of the populace. Hampshire could be thriving while Yorkshire writhed in the toils of famine. News took days or weeks to arrive, and news from America could take months. The sources of power were three: wind, which drove ships and gristmills; water, which also drove gristmills; and the muscles of man and domesticated beasts. It had always been that way, and although the new liberated science might dream of discovering the secrets of nature while enlightened philosophy pondered the perfectibility of man, to the common man the world always would be that way. And on January 17, 1706, in the colonial city of Boston, it was into this world that Benjamin Franklin was born.

CHAPTER ONE

FRANKLIN, THE ADROIT OLD ADVENTURER

Let the experiment begin.

— *Benjamin Franklin*

In something resembling a state of pique, Benjamin Franklin, that most well regulated and calmest of men, decided to go fly a kite.

Although the event has become deeply embedded in American folklore, opinion is still divided on whether this, his most famous experiment, actually took place. His own description of the 1752 event in the *Autobiography* is uncharacteristically vague, an unusual lapse for one of the clearest stylists and sharpest observers in the English language. The full account remained unwritten for fifteen years after the experiment itself, until in 1767 the English chemist and religious philosopher Joseph Priestley published *The History and Present State of Electricity*, a two-volume work heavily influenced by Franklin (some believe he wrote all or most of it, because Priestley was a chemist) that remained the definitive study of the subject for the balance of the eighteenth century. But to doubt the existence of the experiment is to quibble in the manner of the early twenty-first century. The event took place on a day in June, in a field outside Philadelphia that was equipped with a shed, and it happened as it did because Franklin was angry.

The tale of the actual kite itself begins in 1750, when Franklin wrote the latest in a series of scientific letters to Peter Collinson, a prosperous London mercer, knowing that Collinson would, as he had always done in the past, pass the correspondence on to his colleagues in the Royal Society for comment and discussion. Franklin had not been to London since 1726, when as a young man he had been stranded there in a doomed attempt to buy the presses and type that he needed to set himself up in the printing trade back home in Pennsylvania. Collinson was an amateur naturalist of some note, who had exploded the popular notion that swallows hibernate in streambeds, and his connection with Franklin was fourfold. He was the London agent of the Library Company of Philadelphia, an organization Franklin had founded to establish the first public library in America. In the course of their library correspondence, Collinson had rekindled Franklin's interest in electricity — some time before, Franklin had bought some apparatus from a traveling electrical showman but had set it aside — by sending him one of the glass rods that, when rubbed, produced sparks and other electrical phenomena in the current London parlor craze for toying with electricity. Because of his scientific bent, Collinson was well informed of new developments in England and on the Continent, and once Franklin began his own series of groundbreaking experiments, he relied on Collinson for word that he was not needlessly duplicating work that had already been done. And because Collinson, a gentleman amateur, was a member of the Royal Society, he was Franklin's conduit to the admired world of European science. Although Franklin had told Collinson to keep the contents of his first electrical letter to himself, Collinson had ignored Franklin's instructions and passed it on. Now, passing on the letters describing his electrical experiments was routine. The subject of the latest, in 1750, was lightning.

As always when he was writing on complex matters, Franklin's text was extremely spare and simple — so simple, in fact, that from the perspective of a later century possessing vastly superior knowledge, to some it might appear comically simple-minded. It was not.

Franklin was proposing to go where no one had ventured in the four thousand years of humankind's recorded history. First, he established his premise: that there were obvious similarities between lightning and the phenomenon he called, following the scientific practice of the time, "the electric fluid," although he soon realized that it was not, in fact, a fluid, despite the fact that it sometimes seemed to behave like one. But he never discovered what it was. In any event, the similarities he listed between lightning and electricity were:

1. Giving light. 2. Colour of the light. 3. Crooked direction. 4. Swift motion. 5. Being conducted by metals. 6. Crack or noise in exploding. 7. Subsisting in water or ice. 8. Rending bodies it passes through. 9. Destroying animals. 10. Melting metals. 11. Firing inflammable substances. 12. Sulphureous smell.[1]

This was a series of simple observations of real-world events, and perhaps he was not the first to notice them. Isaac Newton, among others, had surmised that lightning and electricity were one and the same. Franklin was the first, however, to take the next step. "*Let the experiment begin,*" he wrote in the 1750 letter. "On the Top of some high Tower or Steeple place a Kind of Sentry Box big enough to contain a Man and electrical Stand." An electrical stand, Franklin's own invention, was an insulating layer of glass or paraffin, the wax commonly used to seal the neck of a jar when preserving food, on which the experimenter stood. Glass or paraffin, he had discovered, did not conduct electricity; neither did silk. He continued:

From the Middle of the Stand let an Iron Rod rise, and pass bending out the Door, and then upright 20 or 30 feet, pointed very sharp at the End. If the Electrical Stand be kept clean and dry, a Man standing on it when such Clouds are passing low, might be electrified and afford Sparks, the Rod drawing Fire to him from the Cloud. If any Danger to the Man should be apprehended (tho' I think there would be none) let him stand on the Floor of his Box, and now and then bring near to the Rod, the Loop of a Wire, that has one end fastened to the Leads; he holding it by a Wax-Handle. So the Sparks, if the Rod is electrified, will strike from the Rod to the Wire and not affect him.[2]

It was a long way of saying something very simple: To prevent the lightning from blasting him to his reward, the experimenter would stand on a nonconductive substance while a roof over his head kept off the rain. Lightning would strike the bent metal rod, and Franklin would touch the rod with an insulated device that would have looked to his contemporaries like a rabbit snare. If the task was accomplished, he believed, he would sample the lightning and live. Probably. No one had ever conducted such an experiment, and Franklin had no idea of the gigantic risk he would be running. He could have been electrocuted, and the fact that he never was during his fifteen years of experimentation was a matter of simple luck. He was never hit with a large enough charge to kill him.

Franklin sent his letter to Collinson and awaited the reply. He knew the members of the Royal Society invariably perused his electrical letters with great interest and enthusiasm. But from London came only silence. At a remove of two and a half centuries it is impossible to determine the cause, and it was not very clear at the time. For whatever reason, Franklin's report of the "thunder gust" (from certain of the letter's language) and "sameness of lightning with electricity" (words derived from Franklin's list) seemed to have fallen through the cracks. In a lifetime of careful self-discipline, Franklin had learned to maintain a placid exterior — a quality that infuriated his enemies — but close observers had seen the man go white with anger, and something of the sort happened now. He concluded that he had been "laught at by the Connoisseurs." But Franklin was a notoriously patient man, another of his infuriating traits. And so, confronted with London's silence, he waited quietly in Philadelphia for the artisans to complete the steeple on Christ Church. Indeed, he contributed to the building fund, and he designed a lottery to raise more money and finish the job. When the steeple was done, he would build his sentry box somewhere on it. Then he would perform the experiment.

Meanwhile, Franklin's electrical letters, including the thundergust letter, had been published in England in 1751, and a bad translation reached France, where the scientist Thomas François d'Alibard was intrigued by the "sentry box" experiment. Franklin, an ocean away, had no inkling that anything was happening.

In May 1752 d'Alibard set to work in the Paris suburb of Marly. He did away with the steeple-or-tower arrangement and at ground level erected a forty-foot iron mast tipped with brass. Instead of an electrical stand of paraffin, he erected the rod on a slab of wood with three wine bottles for legs. Then, at the crucial moment, he found himself engaged elsewhere, leaving a former dragoon named Coiffer or Coiffier in charge of the apparatus.

At twenty minutes past two on the afternoon of May 10, a clap of thunder sounded, followed by hail. Coiffer tested the mast with an "electric phial" (whatever that was), drew off sparks, and quickly sent a nearby child to fetch the local prior. Hearing the news, the prior began to run through the falling hail. The villagers ran after the prior, convinced that the brave cavalryman had met his maker. It was, all in all, quite a procession that soon found the cavalryman in fine shape. The prior must have been Franklin's equal in self-possession. With his own hand, the prior drew off the electric fire — apparently he used something to produce a spark — and emerged from the experience unscathed. In other words, he too produced some sparks. Then he quickly scribbled a report and sent it to d'Alibard, who reported complete success to the Royal Academy of Science three days later. All France resounded with praise for Franklin, the man who designed the experiment that drew the fire down from heaven.

In 1752 in Philadelphia, unaware of the French experiment, Franklin was out of patience. More than twenty-four months had passed since the thunder-gust letter, there was still no steeple on Christ Church, the supposed snub from his British colleagues still rankled, and it was time to take matters in hand with an even more elegant, even better designed experiment than the one he had originally planned. A kite, he decided, would give him "better access to the regions of thunder than by any spire whatever."

According to Priestley, Franklin feared ridicule if he failed, so it would not be a public demonstration. But he planned carefully. For the kite itself, he would use silk, because it was "better able to bear the wind of a thunder-gust without tearing." From the top of the kite protruded a foot-long, sharply pointed wire that resembled an aerial. It was a flying lightning rod. The wire, Franklin reasoned, would attract the electricity in the cloud cover, either in the form of lightning or as a current passing from drop to drop of water. Then he made his ground connection. "To the end of the twine [that tethers the kite to the ground]," he wrote afterward, "next the hand [of the kite flier], is to be fastened a silk ribbon [for insulation], and where the silk and twine join, a key may be fastened."[3] The metal key was his electricity detector. He would touch it with something and produce a spark.

In June, just a month after d'Alibard had made Franklin the unwitting toast of the *ancien régime* in France, the weather in Philadelphia looked promising for the purposes of electrical phenomena. It was overcast, threatened rain, and hinted at thunder — and thunder usually meant lightning. With his illegitimate son Billy — a young man of twenty-one and not the stripling depicted in the historical prints — Franklin repaired to a distant field with a convenient shed, whereupon everything seemed to go wrong. The kite was sent aloft, but nothing happened. Then Franklin noticed something. The loose threads of the twine were standing erect and away from each other. Franklin's experiments had taught him that this meant the twine was electrified, although he wouldn't have used the word because the word wasn't invented; Franklin would have said "electrised." He approached the key with his knuckle. A spark leapt across the air between key and knuckle. He had brought a

Leyden jar with him. Quickly he attached it to the kite string and, Priestley wrote in his history of electricity, "he collected electric fire very copiously." And took it home.

Through his usual channel, Franklin reported the experiment to the Royal Society. Collinson's reply assured him that there had been no snub and that any rift, had there been one, was now a thing of the past. In England, still in that same memorable year of 1752, Franklin's friend John Canton successfully performed the sentry-box-and-steeple experiment. In Russia, however, a Swedish scientist named Georg Wilhelm Richmann failed to read Franklin's instructions quite so carefully and became the first fatality of the electrical age when he was struck by what appears to have been ball lightning. Ball lightning is a peculiar phenomenon, a ball of lightning tethered to no cloud, and most experts claim it doesn't exist, although many thousands of people believe they have seen it. The tsar, who had paid a great deal of money for Richmann's services, banned all further electrical experiments.

Yale, Harvard, and William and Mary gave Franklin honorary degrees, as later did Oxford and St. Andrews when he was resident in England as the Pennsylvania colony's emissary to Parliament. He received the Royal Society's Copley Medal, the highest award of the most distinguished scientific body in the world, and became a member. But although the chorus of praise was deafening, it was not universal. Without explanation King George III ordered his science advisor to declare that Franklin's science was all wrong. When the advisor told his royal master that the laws of science could not be changed by royal whim, he was fired.

And in France the abbé Nollet, chief royal electrician, was deeply chagrined and, it appears, slightly unhinged by Franklin's success. Until this provincial buffoon Franklin — who, Nollet believed, might not even exist — had touched the lightning, Nollet had been the foremost electrical scientist in the world. Perhaps Franklin's subsequent invention of the lightning rod for houses also played a role in the abbé's discomfort. Nollet wrote patronizing essays belittling Franklin's science and, when the imperturbable

Franklin did not react with anything resembling outrage, chagrin, bluster, or shame and the scientific community did not look with favor on either Nollet or his publications, Nollet set about a public demonstration designed to prove that Franklin was dead wrong about everything. Apparently it was not a very good demonstration — few details survive — for it was unmasked as a hoax, and little more was heard from Nollet on the subject of Franklin, provincial buffoonery, and Franklin's electrical achievements.

The same could not be said of the rest of the clergy. Thomas Prince, the pastor of Old South Church, laid the blame for the Boston earthquake of 1755 directly at the feet of Franklin's heretical new lightning rods, which he had invented as a result of the thunder-gust experiment. The metal rods, placed on the tops of houses and other buildings, attracted lightning strikes and conveyed the electricity harmlessly into the ground: a structure with a rod was no longer struck by lightning. In the Reverend Mr. Prince's view, Franklin had interfered with the heavenly order, and God had gone into a big snit. "In Boston," thundered the divine, "more [lightning rods] are erected than anywhere else in New England, and Boston seems to be more dreadfully shaken. Oh! There is no getting out of the mighty hand of God."[4]

European churchmen saw the matter somewhat differently. For many centuries, the European church — both Catholic and Protestant — had known perfectly well what caused lightning. Demons did it, a view that enjoyed the endorsement of both Thomas Aquinas and Martin Luther. The solution was to ring church bells. The demon theory, unfortunately, revealed a singular disconnection between received wisdom and observable fact: in Germany, for example, within the span of thirty-three years, some four hundred church towers were damaged by lightning and 120 bell ringers were electrocuted, crushed by debris, blown to atoms, or otherwise sent to their maker. Nonetheless, churchly feelings ran high against the installation of Franklin's lightning rods. In France, Switzerland, and Italy, the pious caused lightning rods to be removed from structures where the enlightened had erected them. This state of affairs persisted until at least 1767, when the Republic of Venice stored a large quantity of gunpowder in the Church of San Nazzaro in Brescia. The rodless church was struck by lightning, the gunpowder went off, one-third of the city was leveled, and three thousand lives were lost, after which little more was heard about the impiety of Mr. Franklin's invention. Still, the old ways died hard. In 1784 a celebrated French legal case was finally settled in favor of a lightning-rod-owning nobleman, whose attorney argued the case for the rod so persuasively, with such a mastery of the principles of science and such enlightened clarity of thought, that the young man's reputation was made. His name was Robespierre.

* * *

By his own account, Franklin's active interest in electricity began in 1743, when he was on a visit to his old home in Boston. There he caught the act of Dr. Archibald Spencer, a traveling lecturer who gave a course, at six pounds a pupil, on the subject of Experimental Philosophy. Among the doctor's bag of tricks was a glass tube that attracted gold leaf — beaten gold sheets the thinness of paper — and brass when rubbed, and a somewhat shopworn demonstration where he suspended a real live boy from silken cords, rubbed the lad's bare feet with one of the electricity-producing glass rods, and drew off electric fire from his face and hands in the usual form of sparks. Soon, perhaps at Franklin's invitation, Spencer was in Philadelphia, where he set up shop in Franklin's post office. This post office was the nerve center of the royal mail network that Franklin, as deputy postmaster general of the colonies, had improved and made to turn a profit. It was his proudest achievement. Franklin was intrigued but not overly impressed by the man or his abilities. The doctor's demonstrations, he wrote, "were imperfectly perform'd, as he was not very expert; but being on a Subject quite new to me, they equally surprised and pleas'd me." Before the doctor departed and shuffled off the pages of history and into his other occupations as Anglican priest and male midwife, Franklin purchased his equipment. And set it aside. Then in 1747, as we've seen, Peter Collinson sent the Library Company a glass rod as a scientific curiosity — the Company collected them — and Franklin's interest was immediately rekindled, to the point that electricity began to consume all his spare time and caused him to execute his planned early retirement from the printing business. He sold half of it to his talented foreman and became a silent partner. For the next ten years, the study of electricity was his business.

At first his new activities seemed little more than an American version of the contemporary craze for electricity among Europe's science-besotted chattering classes. Franklin attempted the suspended-boy experiment. He also attempted to electrocute a turkey, with unfortunate results when he shocked himself in the process. Reporting "a crack as loud as a pistol shot," he wrote that "I felt I know not well how to describe, a universal blow throughout my whole body from head to foot, which seemed within as well as without; after which the first thing I took notice of was a violent quick shaking of my body, which gradually remitting, my senses as gradually returned."[5] The turkey survived. Franklin had missed his mark and hit himself, and apparently he didn't have the heart to continue, but he had discovered that turkeys, unlike chickens, were hard to kill. He had to hit them with a real big jolt. And, inadvertently, he had just discovered that people were even harder to kill than turkeys. Undaunted, he wrote to Collinson in London and requested more electrical devices; Collinson sent him another glass rod, about three feet long and three inches thick. Franklin also began to build devices out of household implements. In one experiment he directed:

> Place an Iron Shot of three or four Inches Diameter on the Mouth of a clean dry Glass Bottle. By a fine silken Thread from the Ceiling, right over the Mouth of the Bottle, suspend a small Cork Ball, about the Bigness of a Marble; the Thread of such a Length, as that the Cork Ball may rest against the Side of the Shot. Electrify the Shot, and the Ball will be repelled to the Distance of 4 to

5 Inches, more or less according to the Quantity of Electricity.[6]

Franklin had demonstrated the mutual repulsion of positive and negative poles of electricity — that is, a positive pole repels a positive and a negative repels a negative — a repulsion that, like its cousin magnetism, occurred invisibly unless a scientist, called in those days a natural philosopher, could make it visible, as Franklin did. Today, it is an experiment performed by children. Then, it was cutting-edge science.

In an age that revered the "new science" of William Gilbert and Francis Bacon while still having a profound interest in magic, and whose greatest scientist, Isaac Newton, was secretly an alchemist, electricity, especially the lightning rod, was the making of Benjamin Franklin. Without it, he would probably have been ranked among the nation's founding fathers as somewhere above Gouverneur Morris and C. C. Pinckney and somewhere below Thomas Jefferson, James Madison, and George Washington, a boon to minor biographers and generations of paper-writing graduate students but a man unknown to the general public.

All his working life, except at the beginning of it when he worked in his father's Boston tallow works, he was a printer, and he called himself a printer to the end of his days. It was not quite the job it is now. For one thing, it involved muscle, and well into middle age Franklin was an unusually strong man. For another, it involved as a matter of course certain editorial functions, as the printer improved the client's prose and thinking. Franklin also published and wrote for the most influential newspaper in Pennsylvania. He annually published *Poor Richard's Almanac*, an astrological catalog, one among many that purported to foretell the future while treating their readers to bits of homespun philosophy — "Little strokes fell great oaks" — that both reflected and shaped the intense practicality that was becoming an American national characteristic. He published, and perhaps drew, the first American political cartoon. He possessed, arguably, the first modern sense of humor, deadpan, dry, and based on exaggeration, and in the *Autobiography* he wrote what some have called the first American book, a tale narrated in a distinctly American voice that was interested in the way things worked and how much they cost. He organized the first American public library, the first volunteer fire department in Philadelphia, and he founded the academy that later became the University of Pennsylvania. When he was the colonial deputy post-master general, he shared the title with a colleague, a Virginian, but Franklin did most of the work. He was a member of the colonial legislature, a colonel and briefly commander-in-chief of the colonial militia, and the first governor — then called president — of independent Pennsylvania after the Revolution. He attended the Continental Congress in 1776 in Philadelphia, where he edited Thomas Jefferson's Declaration of Independence, including the crucial and deathless opening sentence. He was American minister to France during the Revolution, and he raised the money that won that Revolution.

In Europe, he made certain august persons nervous. The king of France put his likeness on the bottom of a chamber pot. To his face, the Holy Roman emperor Joseph II confessed that Franklin made him uneasy because, the emperor wittily remarked, "I am by trade a king." He attended the Constitutional Convention in 1787 and made valuable contributions, including the convention's punch line when a matron asked him what he and his fellow commissioners had given the country: "A republic," he said, "if you can keep it." He invented and played a once-popular musical instrument called the glass harmonica, whose ethereal tones resembled those of the much later theremin. Mozart and Beethoven composed for it. He invented the Franklin stove and bifocal lenses but, as with the lightning rod, did not patent them although both England and America had patent laws, because a philosopher's inventions were to be freely given to mankind. He was the first to map the Gulf Stream.

But it was electricity that admitted him to the pantheon of the great. Because of electricity, he became the most famous person in America. Because of electricity, he was for a time the most famous subject of King George III, until the Declaration of Independence in 1776 made him a subject of no king at all. In the eyes of many, he came to embody American democracy, and some believed that he was the author of it, although for most of his life, until the storm clouds of the Revolution began to gather, he was an enthusiastic British Empire loyalist. He did not, of course, write the Constitution, but people thought he did. Because of electricity and his disciplined, agreeable personality, he became, perhaps, the most popular person in the world. But in the end and in his view, electricity failed him, and he abandoned it.

His research had everything to do with the Scientific Revolution of the seventeenth century, and it had nothing to do with it. The Scientific Revolution had given the world the telescope, the thermometer, and the microscope. William Harvey had discovered the circulation of blood. Newton invented the mathematical discipline called calculus, founded the science of optics, and pondered gravity in a way no one else had ever done. No longer were the events of the natural world ascribed to superstitious wonders and the hand of God. The new science, following the queen's physician William Gilbert, also did not seek its knowledge in the writings of the ancients, and it did not think that everything was known. Instead, it relied on hands-on experiments, like the one Franklin made with the kite. Through experiments, the scientist could examine a thing, discover what it was, and investigate how it worked. Through experimental science, the causes of events like the weather caused by Franklin's Gulf Stream became knowable, they could be described, and they could be tamed for useful purposes. Indeed, taming them for useful purposes was the entire job of the science described by Francis Bacon. Through experimental science, the world could be controlled. This was Franklin's intent as well. With the new experimental discipline of the Scientific Revolution, electricity could be explored and understood, at least up to a point. Then perhaps it could be harnessed. But the new Scientific

Revolution, except its emphasis on experimentation, was of absolutely no use to Franklin as he went forward. Almost nobody in the world knew more about electricity than what William Gilbert had done, and most people didn't even know that; the magnetic nature of the earth was not exactly common knowledge. A handful of people had begun to study it and its sister force, electricity, as Newton had, but electricity gave up its secrets slowly. As Franklin discovered, in the study of electricity there was a level playing field where anyone with intelligence, skill, and the proper instruments could make major discoveries. Once he had done the necessary reading, he was as deeply steeped in the electrical arts (as the science of electricity came to be called) as anyone on earth. In the popular mind, such as it was, electricity was a tool in the conjurer's bag of party tricks. As Franklin drove forward, using the new scientific method that he had mastered in the middle of his life, he began to make discoveries, and he was the first to make them.

As he progressed in his research during the late 1740s and the 1750s, Franklin not only reported his findings to his peers and invited them to duplicate them, as any modern scientist would, but he also took care to identify his assistants. They were: Philip Syng, a silversmith who, with Yankee ingenuity, invented a lathelike machine that dispensed with the tedious hand rubbing of the leather (in Franklin's case, buckskin) on the glass rods, which was the only known way to electrify them; Thomas Hopkinson, who went on to do original electrical work of his own; and Ebenezer Kinnersley, a Baptist minister who also did original work on electricity and became a talented lecturer on the subject. Together, Franklin and his team reached a point where, building on the results of past experiments, the results of a new experiment could be predicted — a sign that genuine science was being done.

Franklin reported to Collinson in 1747 that electricity was "a common Element" — that is, a single phenomenon — and not, as had previously been believed, two separate but similar forms of fluid: a resinous form, residing in amber, and a vitreous form, resident in glass. Franklin arrived at this assessment by rejecting the

conventional wisdom, keeping an open mind, and drawing a conclusion from what he actually saw. And he discovered more. When
electricity was passed from one body to another, Franklin found, it
was not destroyed. The electricity was not missing; rather, it had a
different polarity, a characteristic still imperfectly understood. Thus
Franklin, in his very first electrical letter to Collinson in 1747: "We
say B (and other Bodies alike circumstanced) are electrised *positively*; A *negatively*: Or rather B is electrised *plus* and A *minus*. And
we daily in our Experiments electrise Bodies *plus* or *minus* as we
think proper."[7] Franklin was the first to adorn the electrical poles
with their now-commonplace symbols, and the language he chose is
the language we use today.

And here, in this new language and still in the first letter in
1747, Franklin moved on to postulate one of the central laws of
physics, the conservation of electrical charge, which means that
electricity does not vanish as it moves from pole to pole. This real-
ization, wrote the Nobelist in physics Robert A. Millikan, was
"probably the most fundamental thing ever done in the field of electricity." Franklin also discovered the phenomenon of the induced
charge, the appearance of a positive charge in a grounded conductor when a negative charge is brought near it. From this, too, many
discoveries would flow, although not directly from Franklin's hand.
The more brilliant his discoveries and innovations became, the
thinner his patience grew.

Franklin, who had reached a point where he could predict the
probable results of his next experiment, was far from flying blind,
but he had gotten to a place where he had no words to describe
what he was doing. "I feel a Want of Terms here," he wrote
Collinson, "and doubt much, whether I shall be able to make this
intelligible." He did his best; it is to Franklin that we owe the words
battery, *brush*, *condenser*, *charge*, *discharge*, and *conductor*.

Feeling his way while inventing a new vocabulary, he finally
began to get ahead of himself. On July 29, 1750, Franklin dispatched to Collinson a summary titled *Opinions and Conjectures,
concerning the Properties and Effects of the Electrical Matter, Arising from*

Experiments and Observations, Made at Philadelphia, 1749. In it, drawing on his discoveries, he made ten points that caused Robert Millikan to later declare him the discoverer of the electron, the atomic particle not identified until 1897.

By 1750 Franklin was almost done with electricity; only the kite experiment and the lightning rod remained in his future, and then he was more or less finished with it. He journeyed to London as Pennsylvania's agent — lobbyist, really — in the mother country, carting his electrical apparatus with him, including something called the Philadelphia Machine that generated a spark nine inches long, something never before seen in London. With brief interruptions in 1762 and 1763, when he returned to Philadelphia, he stayed in London for more than twenty years. Then home beckoned again, the Revolution followed, and Benjamin Franklin, deceptively amiable and iron of will, a man who had thought of himself as an Englishman for most of his long life, became one of the founders of a new country. He made no more electrical discoveries. It is tempting to say that he was distracted by events, but this does not tell the whole story. Even in old age, Franklin was a man of prodigious energy; the most casual look at his career reveals that he was entirely capable of doing more than one thing — and doing them well — at the same time. The truth was, electricity had disappointed him.

He was, after all, the founder of the American Philosophical Society for Useful Knowledge — an organization identified by the last three words as a very American organization indeed — an organization dedicated to the principles of the Scientific Revolution. Francis Bacon, in describing the new science for which he was the spokesman, had stressed that its goal was to discover things — useful knowledge, but also useful objects — that would benefit the human race. Franklin was a great believer in this. But his long years in London taught him that Americans were becoming very different from their British forebears. Americans lived longer than the inhabitants of the mother country. They were, especially in the colonies inhabited by the dissenting Protestants, Quakers, Puritans, and Presbyterians, all of whom required their members to seek their salvations by reading their Bibles, vastly more literate. Unlike the English, most of them owned their own land. And Americans, much more than the English, were a deeply practical people, enamored of results and disdainful of theory. "The combination of a *theory* which may satisfy the learned," Thomas Jefferson later wrote, "with a *practice* intelligible to the most unlettered laborer, will be acceptable to the two most useful classes of society."[8] But in a land with few theorists and many unlettered laborers, the natural bias of education was in the direction of practice — the "useful arts," in a phrase coined by Jacob Bigelow, professor of *materia medica* and first Rumford Professor "of the application of science to the art of living" at Harvard College.

To Franklin, that which was not useful was vile, and electricity was not proving to be useful. The lightning rod was a passive device for drawing off a charge, but it was the great and sole exception. One modern observer would write: "Bacon and Descartes predicted that any basic scientific knowledge would lead to practical innovations of use to people everywhere, [but] by 1787 there had been only one spectacular example of disinterested or basic research that had led to an important invention. I refer here to the lightning rod."[9]

The inventor-mechanic had a singular advantage. The world of the Anglo-American inventor-scientist was a remarkably small one, and it was highly accessible because so little was known that no one had much of an edge. New information from whatever source was prized. And there weren't many people who could understand what Franklin was up to. Given the constraints of travel, it took some time for his electrical papers to circulate, but within a couple of years his findings were in the hands of the few people who could fathom what he was trying to do — and carry his work to the next level. The small world of science also meant that word of a new discovery eventually reached those who knew how to shape it into forms that were very new and sometimes strange. It was an information society that Franklin had entered with electricity, a place where people passed knowledge to each other — a seemingly modern phenomenon that is actually very old. This was a gigantic breakthrough. For millennia, nobody had ever been able to figure out what to do with electricity.

But as breakthroughs went, it was not Franklin who made it, at least in his own eyes. In Franklin's view, while the lightning rod *did* something, it *made* nothing — and Franklin, the practical American disciple of Francis Bacon's useful science, was a man who badly wanted to make things. Electricity utterly baffled him. Indeed, the more he learned about it, the more the nature of electricity floored him. He was doing pure science, discovering the laws, principles, and nature of electricity, and pure science could not sustain Franklin's interest for long; he wanted to discover the utility of electricity, and it eluded him. "If there is no other Use discover'd of Electricity," he wrote in one of his letters to Collinson, "this, however, is some thing considerable, that it may *help to make a vain man humble*."

He tried to pass it off lightly.

"Chagrinned a little that we have hitherto been able to discover nothing in this way of use to mankind," he wrote Collinson, "and the hot weather coming on, when electrical experiments are not so agreeable, 'tis proposed to put an end to them for this season somewhat humorously in a party of pleasure on the banks of the Schuylkill (where spirits are at the same time to be fired by a spark sent from side to side through the river). A turkey is to be killed for our dinner by the electrical shock, and roasted by the electrical jack, before a fire kindled by the electrified bottle, when the healths of all the famous electricians in England, France, and Germany are to be drank in electrified bumpers under the discharge of guns from the electrical battery. An electrified bumper," he added, "is a small thin glass tumbler, near filled with water and electrified. This when brought to the lips gives a shock, if the party be close shaved and does not breathe on the liquor."[10] But the revelry could not disguise his chagrin. Try as he would, he couldn't make electricity *do* anything but kill birds and perform party tricks.

Moreover, he'd reached a dead end even in the pure science of laws, principles, and nature. Although he would continue his experiments until he left London for Philadelphia in 1775, Franklin had discovered virtually everything an extraordinary man, using the tools of his time, could learn about electricity. Although he'd devised a lead-foil battery that promised great things if properly modified and developed, he didn't seem to realize its possibilities. Fifty years before anyone else, he'd built a primitive electric motor: a wheel that he turned with electricity. But he lacked a reliable, long-lasting power source that might have taken him to the next level, and his legendary patience was at an end. And so the experiments ceased. He never appreciated the importance of what he had done.

If it is possible to fix a date for the beginning of the Industrial Revolution, it is probably 1769, the year Richard Arkwright built his first cotton-spinning mill and James Watt repaired a coal-fired pump by fitting it with a condenser and thus invented the modern steam engine. But Franklin discovered the first and most basic law of electricity in 1747 and reached his scientific apogee in 1752 with the lightning rod, thus laying the foundations of the Electric Revolution. In the numbers game, then, the Electric Revolution wins the race with its industrial counterpart, thanks to Franklin, because the Electric Revolution began almost twenty years earlier. But it *is* a game, and the important thing is what happened next. Arkwright's mill was immediately copied, and Lancashire became the cotton cloth manufacturing center of the entire world. Watt and his partner, the Birmingham manufacturer Matthew Boulton, had their steam engine works up and running by 1775. But it took until the 1830s, almost sixty years, before the Electric Revolution bore its first practical fruits. They were two. In the interval, the electrical scene shifted from America to Europe. It involved a lab accident, two Italian professors, both of Charles Darwin's grandfathers, an American renegade, and a former apothecary's apprentice. Only then was the world of electricity ready to move on.

CHAPTER TWO

ERASMUS DARWIN, HUMPHRY DAVY, AND THE SCIENCE OF THEIR TIMES

In America, Franklin had no real successor after dying at a good age in 1790. The electrical scene moved to Europe, where, using electricity, the Englishman Humphry Davy redefined the science of his time. And to understand what that science was, we must take a step several decades back and direct our gaze to the figure of a corpulent libertine who played the trombone to his flowers.

The corpulent libertine was Dr. Erasmus Darwin, grandfather of the great naturalist and principal physician in the beautiful cathedral city of Lichfield, Staffordshire, a man so eminent he turned down the king's request to become his personal doctor and got away with it. Dr. Darwin was intimately familiar with most of the science of the late eighteenth century, and he knew where to find the rest of it. Once a month in Lichfield, or in nearby Birmingham on the night of the Monday nearest the full moon, he sat down, broke bread, and hoisted tankards with the founders of the Industrial Revolution and the greatest British chemist of his time. They called themselves the Lunar Society. Success has many fathers, but it appears that Dr. Darwin founded the club, gave it its name, and called its members lunatics. As the eighteenth century waned, the subject of their talk was not electricity. It was James Watt's steam and Joseph Priestley's unphlogistated air — the colorless, odorless, invisible gas Priestley had codiscovered, later given the name of oxygen. Electricity, unless Benjamin Franklin was in town, was something of an afterthought.

In number the members of the Lunar Society were fourteen, though some say thirteen. Many Americans will recognize four: the chemist Priestley himself, who was also the popularizer of Franklin's electrical discoveries and a theologian of such extreme views that even the Unitarians considered him radical; Josiah Wedgwood of crockery fame, who was Charles Darwin's other grandfather and who built the world's first steam-powered factory at Dr. Darwin's urging but may have lived too far away to qualify for full membership; James Watt, the neurotic former instrument maker who repaired a model of a steam-powered pump and inadvertently invented the steam engine; and Matthew Boulton, the Birmingham toy maker — he was also a silversmith, coiner, and manufacturer of shoe buckles and swords — who built Watt's engines and lived in the first house in England since the Romans left to be equipped with central heating. Benjamin Franklin was not a member, although he sometimes traveled and attended meetings, corresponded with the members in his absence, and had planned to retire in England before events intervened.

Central to the proceedings was, of course, Dr. Darwin, by all accounts a most clubbable man. He raised three families with two wives and a mistress, was a gourmand of such determination that he had a semicircle cut from his dining-room table so he could get closer to his food, and anticipated his grandson's theory of evolution, abandoning his studies in that line because, as a doctor, he had to think of his reputation. He invented the steering wheel, discovered photosynthesis, wrote immensely popular and utterly dreadful poetry, dashed about the countryside in a carriage especially designed not to tip when driven at high speed, and was forever pestering Boulton to build him a car powered by steam. In all these things, he represented the last flowering of an age that was now passing under the onslaught of the new science and the twin revolutions of electricity and steam.

In spirit and accomplishment Erasmus Darwin was similar to

Thomas Jefferson, an immensely talented dilettante of the gentlemanly school: Dr. Darwin could boast of some very pretty, even distinguished accomplishments, but — and here he was unlike Jefferson — he was never pressed by events to rise above himself. Darwin was a deeply likeable man, as the regard of his friends no less than the evidence of history attests, but in a later age he would have been described as a person who did not live up to his potential. Although he was capable of thinking well, he did not think deeply; he was too comfortable and too pleasurably engrossed in the present to think too far ahead. To judge by the writings he left behind, doubt was a stranger to him. Moneyed, talented, ingenious, and possessed of an abiding, if selective, curiosity, Dr. Darwin was disinclined to look far beneath the surface of things. In this, he represented a science that was now passing — a slapdash science influenced but not deeply informed by Francis Bacon and too much based on anecdote rather than disciplined experiment, where you sometimes made a good guess, occasionally got lucky, and were often wrong but didn't know it. In modern science, you still get lucky, but almost nothing else is the same. To Dr. Darwin, good company and occasional inventor that he was, science was something to discuss with his friends, a subject of his dreadful poetry, and an entertaining, interesting phenomenon whose alleged facts did not always require close scrutiny. There were many Dr. Darwins in the England of the eighteenth century's last half. It may not go too far to say that the intervention of a remarkable rascal named Benjamin Thompson changed everything.

He was an odd man, Benjamin Thompson, even for a time when many men seem distinctly strange to modern eyes. Born on a farm in Massachusetts, he was run out of Rumford, New Hampshire, for philandering, fought on both sides in the Revolution, and finally stuck with the British because the pay was better. He moved to London, where he rose in society until George III gave him a knighthood to get rid of him, and became the only American except Douglas MacArthur to rule as the absolute dictator of a foreign country. The country was Bavaria, where the elector — as the country's ruler was styled, because he was one of the few people entitled to vote for a new Holy Roman emperor when the office fell vacant — gave Thompson the government, no doubt for what seemed like good reasons. Setting up shop in Munich, he proceeded to invent the drip coffee pot, thermal underwear, the thermos bottle, the kitchen stove, the roasting pan, central heating, the welfare state, the modern theory of heat, and possibly even baked Alaska. A brand of baking powder and a remarkably efficient fireplace still bear his name. The elector responded by making him a count of the Holy Roman Empire. Thompson took as his title the name of the New Hampshire town that had thrown him out; he is remembered in history as Count von Rumford. Rumford, New Hampshire, responded by changing its name to Concord.

Returning briefly to England as Count von Rumford before finally settling in France and marrying the widow of the great and beheaded French chemist Antoine Lavoisier for her money, Thompson made his last contribution to the betterment of the human race by establishing the Royal Institution, a combination research lab and lecture hall where the disciplined, cutting-edge science of the day would be on public display and also practiced. It would be, Thompson hoped, the finest research facility in the world. One of his first hires was Humphry Davy, and he got his wish.

Davy, who had begun his working life as an apothecary's apprentice, had become at a very young age chief chemist of the Pneumatic Institution near Bristol, where at Dr. Darwin's urging the patients were administered gallons of oxygen a day. The fact that the therapy didn't work, either as advertised or at all, was obscured by a reliance on anecdotal results that were the warp, woof, and bane of the older science Dr. Darwin represented. But although the eminent doctor remained a keen fan of the therapy, Davy soon concluded that the work of the Pneumatic Institution was based on a forlorn hope. His Bristol endeavors, Davy wrote, were "the dreams of misemployed genius which the light of experiment and observation has never conducted to the truth." Unlike electricity, at least when Davy got around to it, oxygen therapy didn't do anything.

Which is not to say that the misemployed young genius didn't have a certain amount of fun. At the Pneumatic Institution he discovered the virtues of another of Priestley's discoveries, nitrous oxide. He named it laughing gas, became addicted to it, and nearly killed himself. He also introduced it to a friend, the poet Samuel Taylor Coleridge, who didn't need another addiction. But Davy's association with Coleridge is interesting for another reason. Coleridge, deeply influenced by the German philosopher Friedrich Schelling, was a titan of contemporary New Age thinking in the form of Romantic poetry. Davy, for his part, came to be known as the prototype of the Romantic scientist.

At the Royal Institution, he became intrigued by electricity, but it was a very different form of electricity from what Franklin had known. Franklin's electricity was lightning, or it was generated when a glass rod was rubbed with leather, or it could be stored for a short time in a Leyden jar. Now, down in northern Italy, a physicist had learned to store it for a long time. And when electricity could be stored for a long time, a scientist like Davy could fool around and try to figure out what he might do with it.

If Franklin had no immediate successor in America, it could be said with equal justice that he had no successor anywhere in the world until 1791, eleven years after one of the most famous lab accidents in history occurred in an unlikely place, the University of Bologna in Italy, in the presence of an unlikely man, a forty-three year-old professor of anatomy named Luigi Galvani, who suddenly decided he knew what electricity was, although he took his time in publishing his findings. In the interval, he had been fooling around with some dead frogs.

The University of Bologna, as it happened, owned one of the successors to von Guericke's friction generator. In 1780 it happened to be operating in the same room where Galvani was either dissecting a frog or trying to eat one for his dinner (accounts differ). The frog's leg twitched: this was the famous lab accident. The frog's leg, Galvani found, did not twitch when it was touched with a glass rod, but only when it was touched with a metal knife. It didn't seem to twitch if the friction generator wasn't running. In truth, the generator had absolutely nothing to do with the effect, but Galvani didn't yet know that. And at first he failed to notice that the frog's leg didn't twitch unless it was resting on a metal plate.

Like Franklin, Galvani was on to something very basic — nerves transmit the messages from the muscle to the brain with electrical charges — but he didn't recognize it. Instead, he concluded that the frog's leg was somehow picking up the electricity from the friction generator, and to test his theory he hung dead frogs from his garden railing to measure the electricity in the atmosphere during the approach of a thunderstorm. He got a minimal amount of twitching (or so he thought) but not much, began to squeeze the frogs in frustration, and eventually gave the whole thing up as a bad job. "So easy it is to deceive oneself in experimenting," he wrote, "and to think that we have seen and found that which we wish to see and find." He had been trying for atmospheric electricity, and he decided he couldn't get it.

But Galvani was a tireless man in an age of tireless men, and he pressed forward down the corridor of years that led to the publication of his findings in 1791, discovering that the twitching effect appeared only in the presence of two different metals as represented by the original scalpel and the dish on which the frog lay — and not just any metals. Brass and zinc were particularly effective, he found. Moving his work back indoors, he discovered that the two-metal twitching effect could be produced in the absence of an external source of electricity. Galvani believed, both firmly (because the leg twitched) and vaguely (because he had no idea why), that he had discovered the secret of life itself, the invisible substance that was central to all living things. Also a believer was Galvani's nephew, Giovanni Aldini, who won the British Royal Society's Copley Medal when he seemed to bring a criminal's corpse to life by convulsing it with a good electrical shock. The corpse winked, and its limbs moved. This erroneous conclusion and interesting experiment had two results, one distant and one almost immediate. Nineteen-year-old Mary Shelley, writing her first book in Geneva in 1816, equipped Victor Frankenstein's laboratory with electrical apparatus and brought his stitched-together and very dead monster to life with a resounding jolt of the stuff. And, closer to Galvani's own time, Alessandro Volta, a physics professor at the University of Pavia across the neck of Italy, became intrigued with his colleague's findings. He duplicated Galvani's experiments, concluded that the frogs themselves weren't generating the electricity, and then noticed something else: neither were the two metals. Nothing occurred when he touched the two metals together. Therefore, something else had to be happening. Clearly the frog, or some substitute for the frog, was causing the twitching. The substitute for the frog's legs he originally chose was his own tongue. When touched by Galvani's two metals, Volta's tongue definitely detected some electricity.

In 1796 Volta stacked together alternating strips of zinc and brass, separated them with cardboard soaked in brine, his chosen frog and tongue substitute, and produced an electrical spark. Next he filled a series of nonmetallic cups with brine, which seemed to be a better conductor than water, placed a strip of zinc in one and a strip of brass in another, connected them with a wire, and produced another spark. The output, he discovered, was far weaker than a Leyden jar's, but Volta's device didn't need to be recharged from an outside source. It would continue to produce electricity until the metal corroded. A Leyden jar sometimes produced a lot of electricity and sometimes very little, and its user never knew when. Using Franklin's term, Volta called his gadget a battery. The world, which thought a battery was four pieces of artillery, called it a Volta pile. Napoleon, the temporary ruler of Italy, gave Volta a medal.

It is difficult to overestimate the importance of Volta's invention. In a world dependent on an unreliable Leyden jar recharged with a von Guericke generator, no practical electrical device — a device, that is, that *used* electricity, Franklin's ever-elusive goal —

had been built. Now, with the Volta battery in hand — in its final form, which was completed in 1800 — Franklin's stalemate was broken and the world would begin to produce a myriad of electrical devices with increasing speed, inventing them in an entirely new way. The Industrial Revolution, just hitting its stride as Volta made his discovery, was about engineering and combustion — finding more efficient ways of burning things, and then finding more efficient ways of applying the resulting force. Like most of humanity's inventive spurts, it was a self-limiting phenomenon.

There was an extremely basic, easily discoverable difference between the Industrial Revolution and the nascent Electric Revolution. The former was based on an abundant, powerful, cheap, and familiar fuel — coal. Electricity, whatever it was, was just the opposite. It was not powerful, it was not abundant, and it was not cheap. But it was strangely mutable.

The Industrial Revolution was not mutable. Once you've invented a locomotive, the only thing you can do is improve it; the invention of a locomotive does not lead to the invention of something else that is entirely new and different. The same is true of other purely mechanical devices. For example, once you've invented a hydraulic brake, it is possible to invent an antilock brake, but the invention of a brake teaches you absolutely nothing about inventing an antigravity belt that allows you to zip around in the air.

According to some scholarly theories, most periods of such inventive creativity last for about seventy-five years. The Chinese, whose technology was equal or superior to Western Europe's in 1500 — and whose groundbreaking devices included the magnetic compass, moveable type, paper money, and the horse collar, the part of the tackle that prevents the strangling of an animal pulling a plow or wagon — owed their remarkable inventory of inventions to the fact that their country had been unified, stable, and well governed for much of its history. But the Electric Revolution that Volta's battery began in the year 1800, when he presented it to the world . . . well, this was something entirely new, and for the moment it was entirely European.

For one thing, the Electric Revolution proceeded with extraordinary speed. The Industrial Revolution did not. Hero of Alexandria had described a primitive steam engine well before the birth of Christ, but it wasn't until the eighteenth century that someone figured out how to *do* something with it — first to pump out the mines of Cornwall, because a piston moved by steam creates a vacuum, and water is drawn into a vacuum, and then as motive power, to create the Industrial Revolution. One hundred fifty years later, the nuclear power plant was born. Like James Watt's engine, it boiled water and produced steam. The nuclear plant was a logical extension of the Industrial Revolution, it didn't do anything different, and it took forever to invent.

By contrast, only thirty-seven years (1800–1837) separated Volta's battery from the inventors of the electromagnetic telegraph. Another twenty years passed before Cyrus Field connected Europe and America with his undersea cable, and twenty years after that Thomas Edison was at work on his light bulb and electrical power stations. Guglielmo Marconi and his primitive wireless telegraph followed by another twenty years, and Reg Fessenden's radio — modern radio — appeared less than ten years later. Compared to the former pace of human innovation, and compared even to the Industrial Revolution when invention seemed to follow on the heels of invention, the Electric Revolution proceeded at a blinding pace. And it continues to accelerate. Today the product cycle of a computer, the Electric Revolution's latest manifestation, is eighteen months. But Volta's battery did not allow the human race to put on such a blinding turn of speed. With Volta's battery great things were done, but sedately. Like most gifts, it came with some drawbacks. As we've said, it was a reliable power source but a weak one. And the electricity it produced was almost prohibitively expensive.

In 1850 an English physicist named Robert Hunt calculated that electric power was twenty-five times more expensive than steam power — and with continued improvement, the price of steam power was falling steadily. The laws of physics could not be repealed, and with the technology of the time, the cost-to-power ratio could not be defeated; a grain of zinc produced only about one-eighth the power of a grain of coal. This drawback was not apparent to the early pioneers of the Electric Revolution. They didn't realize that they had to invent devices that ran on weak electricity. They discovered it by building things that were underpowered and too expensive. It was not until 1882 that an abundant source of inexpensive electricity was built. Until then, if electricity was going to make any money for its hopeful possessors, they were going to have to figure out how to do something useful with a trickle of power.

Because of the way human beings measure time — in terms of their own lifespan — the Electric Revolution seemed to be moving rather slowly. An electrical device would be introduced, people would have a score of years to get used to it, and the next electrical miracle would appear. But if the early pace of the Electric Revolution seemed to be a stately one, just ambling along, it was also true that nothing in human history had ever moved so rapidly. In the world where the Electric Revolution was born, the output of an economy could take more than seven hundred years to double, but when the Electric Revolution began to hit its stride, output could double in half a human life — or less. Nothing had ever gone this fast before. Unprecedented abundance resulted.

And there was one more thing about the Electric Revolution that was without precedent. Only a few years after the introduction of the telegraph, the great cities of America were wired for electromagnetic fire alarms. A telegraph and the fire alarm do similar things, but they do not do the same thing. Edison's large-scale electrical generators bear a certain resemblance to Volta's batteries. Both generate power. But an Edison generator, by even the most generous interpretation, is not the same thing as a Volta battery. A vacuum tube resembles a light bulb: both are encased in glass, and both plug into sockets. But a vacuum tube does something entirely different from a light bulb, although the light bulb inspired it. This was the Electric Revolution, and the man who kicked it off — who brought it into the lives of ordinary people — was Humphry Davy.

* * *

Galvani and Volta had broken Franklin's stalemate — that everything that could be discovered about electricity had already been discovered — and in Europe the march of science had moved the study of electricity to the next level, where it became the temporary province of ex-bootmakers and former bookbinder's apprentices. It was a fascinating interval, and the scene of its unfolding was in England on the stage of London's Royal Institution, where young Humphry Davy, an itinerant chemist of no particular distinction, became the Elvis of his day. Crowds, huge for the times, flocked to hear him, women and girls wrote him poetry, and Mary Shelley stole his laboratory and gave it to Victor Frankenstein.

At the Royal Institution, Davy built the largest Volta battery in the world — two thousand cells, as measured by the number of paired metals submerged in brine. With it, he revolutionized the science of chemistry. Then he invented the electric light. Standing as he did at a crossroads where chemistry, physics, and electricity met, Davy was a pivotal figure — a man in whom many things came together and from whom many emanated. He transformed the European science of his day even as he embodied it.

Examining his new two-thousand-double-plate Volta battery, he concluded that an interaction of the two metals did not produce the electricity. Rather, the electricity was produced by the chemical deterioration of the metals; an electrochemical phenomenon — a force — was involved. This was something Volta had not explored. Now, using the very battery that had given him the insight into an entirely new process, Davy decided to find out where electricity might lead him when it was applied to chemistry. To Davy, the science of electricity as practiced by a Romantic scientist was a form of priesthood, and admission to the shrine was restricted. As befitted a true Romantic scientist, he was the lone hero of his work, the original self-made man. He was also, although he didn't know it, the prototype of the new nineteenth century's emblematic figure, the lone inventor.

* * *

At the Royal Institution, the young Romantic scientist's lectures were soon the hottest ticket in London. These lectures were not for everyone. Thompson, who leaned toward the principles of the Enlightenment, had provided the Royal Institution with a special entrance for the poor. Davy bricked it up. Rumford had erected a special balcony for the lower orders. Davy tore it out. Rumford had intended the Institution to be free to all. Davy charged admission and lectured to a house packed with "the right sort of people." Extremely handsome in an androgynous way, and a spellbinding speaker, he made society belles swoon.

Shortly after the Leyden jar made its appearance in 1745, it was discovered by the Englishmen Nicholson and Carlisle that electricity could decompose water into its constituent elements, hydrogen and oxygen. Using his huge Volta battery and following the same principle, Davy was the first to isolate barium and sodium — which nobody had ever done before because nobody but Volta had possessed a Volta battery, and Volta was not a chemist — and he later correctly identified Element X as iodine. His method, arrived at in 1807 by trial and error, was simplicity itself. It was suspected that the so-called alkali metals such as potash were actually compounds of something. If Davy could somehow break them down into their component parts, he could both analyze them and identify the elements of which they consisted, because by definition an element is something that can't be further broken down. In his first attempts, he tried passing the current from his Volta pile through alkali metals dissolved in water, but this produced nothing but hydrogen gas. Next, he heated the alkalis and made them molten, and again he applied the electricity. In his first success, he extracted potassium from potash, threw the particles into some water, and, said his brother, Dr. John Davy, "danced around and was delirious with joy" as the potassium "skimmed about excitedly with a hissing sound, and soon burned with a lovely lavender light." Next, he separated sodium from salt.

But the Romantic scientist, searching for nature's unifying

force, did not confine himself to a single discipline. In 1802, using the most powerful Volta battery the world had ever seen, he caused a filament of platinum to glow. It wasn't much of an electric light — it didn't last very long, and it wasn't very bright — but it was an electric light all the same. Then in 1809 he astounded his audiences by passing a brilliant arc of electricity between two electrodes made of charcoal. This was the arc light, the century's first form of electrical illumination, and eventually a practical one, although Davy didn't invent it and never claimed to have. Volta was the first to observe spark discharges between the electrodes of his battery, and by 1802, when Davy first informed his audience at the Royal Institution that something like an arc light was possible, researchers were intensely studying the phenomenon in France and the Germanies, that congeries of kingdoms, princedoms, electorates, and free cities that then occupied the territory of the modern nation-state. When a column of gas was placed between two cross-sparking electrodes, it was found that the electrical conductivity of the gas was improved. But for complicated reasons, including the poor conductivity of the air, a spark discharge between two metal electrodes was weak, and it was probably Davy, the man who studied the elements, who came up with the idea of trying charcoal, which is made of the element carbon. He also found, he told his audience, that the resulting arc of electricity was "of a vivid whiteness." Then he demonstrated it.

With the arc light, although it was dirty and it stank — because it produced the light by burning the carbon, and burning carbon is both dirty and malodorous — Davy had discovered a cheap and easily manufactured form of electrical illumination, but he also encountered the problem that was the later undoing of virtually all the early electrical inventors who turned their efforts toward producing consumer products: there simply wasn't enough electricity in the world to make them practical. The Volta battery and its weak-acid successors lacked the power to illuminate a single street, and they couldn't run a single lamp for very long. Until the Russians and Americans solved the problem in the late 1870s, arc light-

ing was principally employed to create theatrical effects. The first
use of Davy's invention came in 1846, when the new Paris Opera
used it to light the skating scene in Meyerbeer's *The Prophet*. In previous theatrical lighting effects, the arc light had been beaten out
by the chemical reaction that produced limelight — by slowly
burning lime — which had the virtue of familiarity. But even in the
theater, the Volta barrier still remained. The Volta barrier was weak
electricity in small amounts.

With his battery, Davy went on to discover boron and calcium.
Without it, he explained the bleaching action of chlorine. He revolutionized the leather tanning industry and wrote the definitive
book on the subject. He saw to it that Sir Stamford Raffles was appointed the first head of the Regent's Park zoo, where Sir Stamford
created the prototype for zoological gardens everywhere. He invented the miner's safety lamp. He became president of the Royal
Society.

But like Carême, the premier French chef of the day, he had (as
Carême said of himself) inhaled too much the smoke of his char-
coals; he had breathed too many dangerous gasses. He never finished his finest work, *The Elements of Chemical Philosophy*, which
according to the Swedish scientist J. J. Berzelius would have "advanced the science of chemistry a full century." He tried and failed
to find an electrochemical way to prevent the corrosion of a ship's
copper sheathing. The experience caused him to proclaim that he
was burned out, and he was right. He wrote a book on fly fishing and
settled in Rome — "a ruin," he said, "amongst ruins." He died in
1830. At the Royal Institution his assistant, Michael Faraday, took
his place. And posterity concluded that Faraday, whom Davy twice
blocked for membership in the Royal Society, was his greatest contribution to science. For Faraday and his American friend and colleague Joseph Henry would take the science of electricity to the
next level, where it was finally possible, both immediately and after
yet another gap of half a century, to produce consumer products
that ran on electricity. Just as they were doing so, however, there
was a curious interval. In 1834, fifty-four years after the death of

Franklin, a Vermont blacksmith built a modern electric motor and powered it with a Volta battery.

* * *

Volta released his battery on an unsuspecting world just two hundred years ago, and the pace of invention has yet to flag. Moreover, the march of electrical science has proved impervious to six hair-raising economic depressions and dozens of recessions. In hard economic times, mechanical invention likewise continues, but nobody has the money to buy the product; development of improved models slows when it does not halt. But electrical invention continues to thrive in good times and bad. More radios, and more new kinds of radios, were sold in the bleak decade that followed the Great Crash of 1929 than were sold in the decade of prosperity that preceded it. The computer revolution made it to the marketplace just as the greatest recession since the 1930s was raging, and it hit its stride just as the bottom fell out of the economy in the early 1990s. In all eras and all decades, the phenomenon persists. It seems like one of God's little jokes that the whole thing began with the leg of a dead frog.

CHAPTER THREE

THE BLACKSMITH OF BRANDON

I sell here, sir, what all the world desires to have — POWER.
— *Matthew Boulton to James Boswell, 1776*

In the year 1833, in the iron country of northern New York, it was a sight unlike any the young blacksmith had ever imagined, much less seen: a three-pound electromagnet holding a 150-pound anvil "suspended," Thomas Davenport, the blacksmith, later wrote, "between heaven and earth." Gripped by one of only a handful of electromagnets in North America, the anvil had been, miraculously it seemed, lifted off the ground by a crane.

Davenport had heard of Professor Henry's wonderful magnet — it was the talk of the countryside — but the actual sight of this marvelous device filled him with a sudden resolve. There was nothing quite like it anywhere in America, and Davenport had to have it. "Like a flash of lightning," he wrote, "the thought occurred to me that here was a marvelous power. If three pounds of iron and copper wire would suspend 150 pounds in the air, what would a 300-pound magnet suspend!"[1] He would soon dramatically revise his thinking of what it could do.

In the North Country, everybody knew about Joseph Henry. The lakeside village of Port Henry on the shores of Lake Champlain was soon to be named for him, and the facts of his still brief but remarkable life — in 1833 Joseph Henry was only thirty-six years old — and his achievements were also becoming well known among his colleagues. Henry had been born poor, the son of a day laborer in Albany, New York, and he was orphaned early. At four-teen, he was apprenticed to a watchmaker, who found him a dull lad because Henry's thoughts were often elsewhere. Henry later told the story that as a youth he had gained access to a church library through a hole in the floor, and there he had encountered the book that changed his life, George Gregory's *Popular Lectures in Experimental Philosophy, Astronomy, and Chemistry*. "This," he later wrote to his son, "opened to me a new world of thought and enjoyment; fixed my attention upon the study of nature, and caused me to re-solve at the time of reading it that I would immediately devote my-self to the acquisition of knowledge."[2]

Perhaps this burning desire was true, although it smacks more than a little of the usual origin story that began to attach itself to American scientists in the nineteenth century: humble beginnings followed by some sort of book-driven epiphany, and then triumph. Still, there is no reason to doubt Henry's word. As he grew to be-come a broad-faced, handsome young man, his first love was the theater. He was offered a job as a professional actor, but friends talked him out of it, and at the age of twenty-two, old for a student, he enrolled in the tuition-free Albany Academy. He made money on the side by tutoring, among others, the elder Henry James, father of William and Henry, the philosopher and the novelist. He also worked as a surveyor and taught school.

In 1826, at the age of twenty-eight, he became the academy's professor of mathematics and physics. At first it was thought he would make his name from the study of hot air balloons, but like William Gilbert two centuries earlier, he became interested in the study of terrestrial magnetism, Gilbert's old insight that the earth it-self is a magnet, a subject not much followed in America. Terrestrial magnetism led to magnetism in general, and at an extremely fortu-nate time. In 1820, while preparing for a lecture, the Danish scien-tist Hans Christian Ørsted had noticed that the needle of a compass was deflected by a nearby electrical wire. Ørsted duplicated his observation under controlled conditions and drew the correct conclusion: magnetism and electricity were related. Almost immediately on hearing Ørsted's news, the impulsive and brilliant Frenchman André-Marie Ampère created an electromagnetic field — that is, an invisible cluster of electricity with magnetic properties — by reasoning thus: if an electrified wire could move a compass needle, two electrified wires should interact magnetically. And so, he found, they did. Two electrified wires placed parallel to each other attracted each other, and nonparallel wires — wires that, when placed in proximity, resembled the letters T, L, and V — repelled each other. But it was still early in the game and he had no idea why it was happening. The angle of their proximity was not a factor. Ampère intensified the effect by forming the wires into coils, but he was a scientist of diverse interests, and he did not pursue his experiments to their logical conclusion. The logical conclusion was to turn electricity into magnetism, and in England the ex-cobbler William Sturgeon, now a lecturer at the Royal Military College at Addiscombe, wrapped a U-shaped core of soft iron with a few windings of wire, connected the wire to a Volta battery, and created the electromagnet. An electromagnetic reaction occurs, as Sturgeon found, when a weak current from the battery turns the iron into an extremely powerful magnet. The year was 1825. Franklin's dream had come true. Electricity could do something. It could lift things.

In 1827, after seeing one of Sturgeon's electromagnets during a trip to New York City, Henry published his first paper on the subject, and in 1829 he demonstrated his own improved magnet, which was tightly wound four hundred times with thirty-five feet of insulated wire "instead of loosely coiling around it a few feet of wire, as is usually described." In addition to its greater lifting power, the new device also required significantly less electricity thanks to the many windings that also increased the magnetic effect. This was a distinct improvement of the magnet, one perhaps inspired by the fact that Henry was working in a part of the world where the brass and zinc raw materials for a Volta battery were expensive and hard to come by. Henry had stumbled onto the fact that a tightly wrapped electromagnet acted as a device called a coil. A coil, Ampère had been the first to discover, increased the power of the electricity. No one knew why.

It was also in 1829, in a series of deductions beginning with Sturgeon's electronomagnet, that Henry discovered the phenomenon of electromagnetic induction, the production of electricity from magnetism, as well as its reverse, the creation of magnetism by electricity. This was another seminal discovery. Without it, the Electric Revolution was impossible, because without a proper understanding of induction, the creation of an electrical generator, then known as a dynamo, would have been impossible, and the world would have depended on batteries for its electricity. But Henry was slow to publish his results, for his early theatrical training had made him a perfectionist, and he carried a full teaching load. In England, Michael Faraday discovered the identical phenomenon at almost exactly the same time, and he beat Henry into print. The two men soon came to share a deep mutual regard, and credit for the discovery was shared in the end, although Henry has been largely forgotten by his countrymen, while in England Faraday's reputation as the greatest physicist of the age remains green to this day. And it was Henry's electromagnet, an invention that he had not patented — for the solitary and noble Romantic scientist gave his discoveries to mankind for free — that Thomas Davenport, the blacksmith of Brandon, Vermont, was now contemplating with something approaching awe. "I did not then," Henry wrote later in life, "consider it compatible with the dignity of science to confine the benefits which might be derived from it to the exclusive use of any individual." Perhaps so, but he was not above making a dollar or two. He had sold one of his magnets to an ironworks.

Henry had pressed on with his work, refining it, learning how to do new things. For Yale University he had built an electromagnet that could lift 2,086 pounds. It was the most powerful magnet in the world. Henry was now teaching at the College of New Jersey (today known as Princeton), where his academic portfolio had expanded to include chemistry, geology, mineralogy, astronomy, and architecture despite the warning he had given when the school first approached him. "Are you aware of the fact," he wrote, "that I am not a graduate of any college and that I am principally self-educated?" Perhaps not, and it was also true that the magnet the blacksmith saw at the iron mine was little more than a stunt, but Henry had also provided the mine with another magnetic device that brought Franklin's thwarted dream to life. This second device performed useful work.

In the rough accounts that survive, including his uncompleted autobiography, it is by no means clear what brought Davenport to the Penfield Ironworks near Crown Point, New York, in that year of 1833. Perhaps he came to buy iron for his forge in the town of Brandon, Vermont, twenty-five miles away. Perhaps he had simply come to see the electromagnet — many did, and the owners of the works obligingly demonstrated it. With business booming in the iron country of northern New York thanks to the expansion of the new railways, the people at the Penfield works were using Henry's improved version of Sturgeon's magnet to separate high-grade ore from the rock and the lower grades. This was Henry's new device, and it resembled a rotating harrow. Magnetized iron spikes were mounted on a circular framework and drawn through the raw material brought up from the mine, combed through it, and with the magic of magnetism separated out the iron, which was then scraped off the spikes. Eventually, however, the magnetism in the spikes grew weak from use, and Henry's electromagnet, employing the principle of induction he and Faraday had discovered, was used to restore their magnetic fields. And Henry's electromagnet never grew weak and never wore out. Using it to lift the anvil was a crowd-pleasing party trick that had nothing to do with normal operations. And Davenport decided he had to have it.

Davenport persuaded his brother Oliver, a tin peddler, to sell his pots, pans, and cart, and to trade his horse for an inferior one and some cash. With the proceeds, Davenport bought the magnet. From the mine owners' point of view, it was cash in hand, and they knew it would be easy for Henry to make another. Oliver Davenport, a practical man with no vision, and now no cart, no horse, and no stock in trade, hoped to recoup his losses by exhibiting the magnet and charging admission. Thomas Davenport had other ideas. With his wife, Emily, taking careful notes, he took it apart to see how it was made.

* * *

Thomas Davenport, although he had never invented anything in his life, proved to be a man cast in the mold of Franklin's example — a simple mechanic who could grasp scientific principles — and Jefferson's theory of the artisan who improved the world by doing things as a man of action. Like most of the men (and they were all men) who made the Industrial and Electric Revolutions, Davenport was a self-taught amateur. William Sturgeon, the inventor of the electromagnet, had been a bootmaker. Joseph Henry, the only American after whom a unit of electrical measurement is named, had apprenticed as a watchmaker. Michael Faraday, the English physicist whose dazzling series of experiments laid the foundations of modern electrical theory, began his working life in a bookbinder's shop. Thomas Newcomen, who invented the first practical steam engine in 1712, had been a Devonshire blacksmith. His engine was used as a pump and did little other useful work. James Watt, who repaired one of Newcomen's engines in 1763, had been an instrument maker at Glasgow University; Watt's repair job, during which he added a condenser to Newcomen's machine, created the motive power that made the Industrial Revolution possible.

Thomas Davenport was born on a farm in Williamstown, Vermont, on July 9, 1802, one of twelve children, not an unusual number for the day. Young Thomas had a reputation as a great reader, especially of science books, but his father died when he was ten, and he was apprenticed to a blacksmith at the age of fourteen. In return for learning the trade, he received room, board, and six weeks of school every year. In all, he received no more that three years of education in his entire life. Nonetheless, he briefly considered becoming a traveling science lecturer along the line of Franklin's Dr. Spencer.

Nothing came of it. When his term of apprenticeship ended, he set up shop as a smith in the hamlet of Forestdale, town of Brandon, Vermont, married, and prospered in a modest way. He does not appear to have been physically robust, and until he encountered Professor Henry's electromagnet there is no record that Davenport had ever done, thought, or said anything out of the ordinary in his entire life.

The myth of the lone inventor — the self-taught genius laboring in solitude against great odds — would soon grow sturdy roots in American ideology. Edison, a shrewd judge of his countrymen, later saw to it that the myth applied to himself in particular, and Davenport's handful of biographers have dutifully portrayed him in much the same way. In fact, Davenport had plenty of help, starting with his wife, who appears to have been a remarkable woman.

Once he had Henry's magnet apart, he understood its secret. From scratch, he built two more Henry-style magnets, tightly wrapping many coils of wire around a soft iron core. Insulation for one of them was provided by his wife, who supplied the necessary silk by cutting up her wedding gown. How he knew he needed to use silk has been lost to history. Perhaps he learned it from his examination of the Henry magnet. Later, after he read up on the subject, Davenport realized that cotton would do just as well.

Both Henry and Faraday had built tiny electric motors. Henry's was designed as a teaching aid, which he called a "philosophical toy," and Faraday's as a research tool. Neither was designed to perform work, and both operated on the same recently discovered electromagnetic principle: if an electric current from a Volta battery or a Leyden jar was introduced into a magnetic field by means of a rotating iron bar, work was performed — not much work, but just enough, and it was measurable because work was the rotation of the bar. By all indications, Davenport had never given a single thought to electromagnetism in his life, until now. And, thinking about electromagnetism, he made a conceptual leap that he believed was uniquely his own.

At this point, after the passage of a few days, Davenport had only duplicated (twice) an electromagnet invented by somebody else, a feat that was well within the abilities of a skilled and intelligent artisan and not a remarkable achievement. Now he obtained financial backing and a nearby workspace from his neighbor, Orange Smalley, a relative by marriage and a blacksmith and wheelwright, a man who made and repaired wheels, ten years his junior, and the two of them set to work while Emily Davenport once again manned her notebook. By late 1834 they had managed to build a motor of their own.

It is an old truism that all science is eventually going to happen, but only when the conditions are right. James Watt, for example, had triumphantly repaired Newcomen's steam engine in 1769, but it wasn't until 1775 that he and Matthew Boulton, the Birmingham toy maker and silversmith, began to manufacture the torrent of steam-engine-related inventions that would change England and the world. The reasons for the long delay were various, many of them having to do with Watt's perennial lack of money, but the greater part of the problem was a lack of necessary engineering. It wasn't until 1775 that John Wilkinson, Ironmaster, the greatest machine-tool manufacturer in the United Kingdom, designed the machine that bored the tight cylinders for the steam engine's pistons that made Watt's engines function with a high and economically meaningful degree of efficiency. To produce an economically meaningful amount of work from a Watt-designed steam engine, the new science of precision engineering had to be invented. An inefficient Newcomen steam engine had been used to pump water out of the mines of Cornwall. An efficient Watt engine could power other machinery, and the modern factory was born. Next Watt invented the governor that prevented his efficient steam engine from tearing itself to pieces with its own power. And he and Boulton took care to patent their engine, because they proposed to sell it to mankind, not give it away in the name of science. A steam engine is a motor. With the exceptions of Henry's and Faraday's philosophical toys, steam engines were the only motors in the world.

The situation was not the same with the electric motor, because only some of the necessary preconditions were in place. It was possible to build one; Faraday, Henry, and Sturgeon had already done it. But Faraday and Henry, at least, had sensed the limits of their invention in terms of the technology of their times.

Faraday, who had wit as well as brilliance, was the first to admit that any vision of the electrical future was somewhat blinkered. When William Gladstone, the future British prime minister, viewed Faraday's own tiny motor, he is supposed to have asked what good it was. Faraday reportedly answered, "I have no idea, but no doubt you'll find some way to tax it." But Davenport was constrained by no such modesty. Like Watt and Boulton, he set out to harness science in pursuit of some real money. Despite Henry's later warnings, he failed to recognize that there was no point in trying to build a commercial electric motor in the middle of the 1830s. Practical science, as Davenport proved to his own considerable sorrow, can happen only when the conditions are right.

Deploying logic and his newly acquired knowledge of magnetism that was solely and, it turned out, erroneously acquired from disassembling the Henry magnet, Davenport began to build his electric motor by barking up the wrong tree. He mounted a magnetized iron bar (the rotor) on a pivot (the stator) and created a field with his electromagnet. Like Ørsted's compass needle, his iron bar was deflected, and like Henry's philosophical toy it began to rotate, stopping dead when the negative pole of the bar came opposite the positive pole of the electromagnet. Any child who has ever played with a magnet knows the phenomenon that results when opposite poles attract each other: using the magnetic field, they glue themselves together and can be separated only with force. For his motor to perform work, it would have to be attached to a belt or a pulley, and it would have to complete a full rotation — not once, but many times. Instead, the rotor made a half turn and refused to budge. Whatever he had built, it wasn't a motor.

He tried switching off his electromagnet as negative approached positive, hoping that inertia would carry his rotor around until he could reactivate the electromagnetic field, creating momentum, but it was a clumsy solution with no promise. Someone would have to be in constant attendance, turning the electromagnet on and off. Next, indicating that he had begun to do library research that was far from complete, he tried a variant of the method that had been employed by the American inventor Charles Grafton Page. He filled small cups with mercury, which he obviously knew was a conductor of electricity. Then he snipped the wires leading from the battery to the electromagnet and placed the ends in the cups. When the magnetized negative pole of the iron rotor approached the positive pole of the electromagnet, he removed the wires from the cups, breaking the circuit, and then replaced them to complete the rotation. It worked, but not well. He still had to stand by the motor, turning the electromagnet on and off by means of the mercury cups. Davenport pressed on.

"In July of 1834," he wrote, "I succeeded in moving a wheel about seven inches in diameter at the rate of about 30 revolutions a minute. It had four electromagnets, two of which were on the wheel, and two were stationary and were placed near the periphery of the revolving wheel." This device is a little hard for the layman to envision. Unfortunately, Davenport seems not to have left any diagrams. He did describe its functioning:

> The north poles of the revolving magnets attracted the south poles of the stationary ones with sufficient force to move the wheel upon which the magnets revolved, until the poles of both the stationary and revolving magnets became parallel with each other. At this point, the conducting wires from the battery changed their position by the motion of the shaft; the polarity of the stationary magnets was reversed; and, being now north poles, re-

pelled the poles of the revolving magnets that they had
before attracted, thus producing a constant revolution of
the wheel.[3]

In other words, he made the rotor go around, but not very well.

Although the text lacks a certain clarity, it appears that Davenport was trying to regulate the rotation with some sort of cam shaft,
the rotating device that opens and closes the valves in the engine of
the modern automobile, "which," says one latter-day electrical engineer, "did not operate very successfully." He wasn't building an
electric motor. Instead, he was trying to build a magnetic motor.

However the actual rotation was achieved, Davenport's neighbors were not impressed. They believed he was working on a perpetual motion machine, the goal of fools and the lure of charlatans
from time immemorial. Moreover, although the wheel undoubtedly
revolved, it otherwise didn't do much, generating only one-fiftieth
of a horsepower. And to achieve this somewhat-less-than-stellar
performance, Davenport had begun to neglect the true vocation of a
sensible man, his forge and smithy. The machine, his mocking
neighbors said, generated nothing but "mosquito power." His father-
in-law urged him to get back to work, and when he appealed to his
local pastor, Davenport's spiritual advisor informed him that "if this
wonderful power was good for anything, it would have been in use
long ago." Undaunted, Davenport took his motor to Middlebury
College, where he gave a demonstration of the "Davenport-Smalley"
device. Middlebury professor Edward Turner recommended that
Davenport lose no time in applying for a patent, or so Davenport reported in his fragmentary and somewhat selectively remembered
autobiography. Perhaps Professor Turner also gave Davenport a
pointer or two.

Running through the myth of the unlettered American inventor is a persistent theme: that native intuition combines with native
cunning to produce a marvelous device — an electric motor, a practical electromagnetic telegraph, an electric light bulb — out of thin
air and whole cloth. This is far from the truth, although a sort of raw

intuition sometimes played a role, and it is nowhere more apparent than in the case of Thomas Davenport. For months, aided by his wife and Orange Smalley, he had been building precisely the wrong kind of motor, one based on magnetism that revealed no understanding of the interaction of electricity and an electromagnetic field. A highly intelligent child might have conceived the same thing. But now, following his trip to Middlebury College, Davenport achieved a breakthrough that could only have been made possible by consulting the texts — in all likelihood at the suggestion of Professor Turner — of that same pure science that had so frustrated Franklin. The good of pure science, as Davenport discovered, resided in the fact that it told a practical man what to do next, out in the real world of artisans and consumers, a seminal discovery that Davenport was not the last American tinker to make.

"I hope the reader does not believe that inventions are 'lucky hits,'" he dictated in his autobiography. "No! it requires profound study, and the most intense application of the mind. To invent is to read nature's laws."

Up until his momentous visit to Professor Turner at Middlebury, a lucky hit was precisely what Davenport was searching for with his magnetic motor. In January 1835 he finally read up on the relevant natural laws, starting with Professor Benjamin Silliman's *Elements of Chemistry*, "which contained some of Prof. Henry's experiments in electromagnetism, by which I learned for the first time the true names of the instruments and materials which I was using." Silliman, a friend of Henry's, was a former lawyer who became the self-taught professor of chemistry at Yale. He also published the country's principal science magazine, the *American Journal of Science and Arts*, a publication nicknamed *Silliman's Journal*.

His research complete, Davenport built a commutator, a device that foiled the polarity problem that kept stopping the works in his flawed machine. With it, for the first time, he was able to construct a true electric motor. He placed strips of insulation on the shaft of the wheel, hammered out his wires until one end was thin, placed the thin parts in contact with the wheel's shaft, and attached the other ends of the wires to the electromagnet. This was the commutator. When the wheel revolved, the current was interrupted by the insulated strips, its polarity was reversed (as if he had turned off the electromagnet), and rotation continued effortlessly. The time had come to alert the world to the new invention — and perhaps pick up a few more pointers while receiving the encouragement so lacking at home.

And from now on, the new machine would be Davenport's alone. For reasons that were never made very clear, Smalley withdrew from the venture. On his own, Davenport set out a new pilgrimage. On his trip to Middlebury College he also had been given a letter of introduction to a convicted felon named Amos Eaton, and now he decided to use it.

Eaton and his college, Rensselaer Academy (now called Rensselaer Institute) in Troy, New York, were virtually one and the same. In 1835, the year Davenport directed his steps thither, it was said that if you wanted to study geology, you went to London or you went to Troy; no less a figure than Sir Charles Lyell, the foremost geological scholar of the time, deemed the city worthy of two visits, so rich were the surrounding strata. Troy owed its eminence to Amos Eaton.

Eaton, who sprang from what used to be called sturdy Presbyterian stock, was a graduate of Williams College who published his first book, on surveying, in 1800 when he was twenty-four. He studied law under Alexander Hamilton in New York City, befriended Washington Irving, and was admitted to the bar in 1802. Eaton was a cultivated, forward-looking man whose mind was broadly open to new ideas. He was deeply interested in geology and botany and founded a school in Catskill, New York, to teach the latter. Like many Americans of the time, however, he also turned his hand to land speculation, which was his undoing. In 1811, in the complex aftermath of one such land deal, he was sentenced to life in prison for a crime he allegedly committed 145 days after his indictment. He served four years, corresponding regularly with leading scientists and public figures, until he was pardoned by the governor in 1815.

Eaton left the law (and land speculation) behind him and joined his old college, Williams, as a lecturer in mineralogy and botany. He did not stay long. In 1819 he moved to Troy, lectured at the Lyceum of Natural History, and added chemistry and physics to his repertoire when he was named professor at the Castleton Medical Academy in Vermont.

He also attracted the attention of Stephen Van Rensselaer, the largest landowner in New York State and reputedly the richest man in America. Eaton became Van Rensselaer's scientific advisor, and together they established Van Rensselaer's personal institution of higher learning, which opened its doors in January of 1825. "I have founded a school," Van Rensselaer wrote in a letter that scholars believe was actually written by Eaton, "for the purpose of instructing persons in the application of science to the common purposes of life."

The Rensselaer Academy became the sort of place that would have warmed the hearts of Franklin and Jefferson, and Eaton was its presiding genius. He called his teaching method the Rensselaerean Plan. It had five parts:

1. Let the students make practical application of all the science.
2. Let the student always take the place of the teacher in his exercises.
3. Let the students attend to one branch of learning at the same time.
4. Let all the amusements and recreation of students be of a scientific character.
5. Let every student daily criticize those whose exercises he has attended.

Following Rule Two, students delivered lectures, perhaps making a virtue of necessity, since in the beginning the Rensselaer Academy had only two professors, one of whom was Eaton himself. Unlike the common practice elsewhere, the students actually performed experiments rather than simply reading and hearing about them, and women, including those from the Troy Female Academy, were

THE BLACKSMITH OF BRANDON   55

allowed to attend. The separate education of the sexes, Eaton claimed, was a "monkish policy from the dark ages." And Professor Eaton's long obsession with geology was on prominent display.

Eaton received Davenport enthusiastically.

"An obscure blacksmith of Brandon, Vermont," Eaton wrote in a newspaper article shortly after their meeting, "happened accidentally to become acquainted with Professor Henry's discoveries in electromagnetism. Possessing one of those minds which cannot be confined to the limits of a blacksmith shop, nor any shop less than the canopy of heaven, he applied this power — with which Professor Henry astonished the scientific world — to scientific mechanics. He turns three horizontal wheels around 50 times per minute with this power. The wheels and shaft weigh 11 pounds. No chemical or philosophical apparatus can hereafter be considered perfect without it. Whatever may be its fate in mechanics, it will cause the name of Thomas Davenport, the inventor, to accompany that of Professor Henry to the ends of the earth."[4] For all the hyperbole, Eaton was perfectly aware of where Davenport's ideas came from. "I told Mr. Davenport," he wrote to Van Rensselaer, "that no one could aid him without Professor Henry's consent." And so he sent Davenport to Henry.

It was a very small world in which Davenport found himself, in some ways smaller than his native hamlet in Vermont. In the realm of American higher education, particularly as it pertained to the sciences, everyone seemed to know everyone else. Joseph Henry, the greatest American scientist since Franklin, is a telling case in point. As a younger man, he had been the tutor of Stephen Van Rensselaer's children. In 1835, when Davenport arrived on Henry's doorstep in New Jersey with Van Rensselaer's letter of introduction, Henry was at the height of his powers, and he had moved far beyond the comparatively small magnet that had struck awe in the blacksmith's heart at the Penfield works.

He was the ideal man to understand Davenport's motor, because in all the world only Faraday knew as much about electricity, and Henry could also understand Davenport, because Davenport was the kind of man he might have been, a highly intelligent, unlettered artisan from an obscure town in the backwoods. Perhaps for this reason, Henry broke his usual rule and actually sat down with the blacksmith when Davenport appeared. Henry was an ideal man but he was also a wary one. There were too many charlatans and mountebanks out there, all of them peddling the new miracle of electricity, many of them seeking his endorsement, and Henry bestowed his favors carefully. The blacksmith's wonderful electric motor did not, however, produce the same rapturous response it had inspired in Amos Eaton.

"The great importance of the invention of the Brandon blacksmith exists principally in the fertile imagination of Professor Eaton," Henry wrote to Professor Silliman at Yale. Eaton's enthusiasm was sadly misplaced, Henry continued, "calculated to injure the inventor of the machine who appears to be a modest unassuming and ingenious young man. The truth is that there is nothing new in the whole affair. Every part of the machine has been invented before and in fact differs [in] nothing from the first one of the kind which I described in the *Journal*, as you will recollect, about 1831."[5]

Henry's 1831 article had, in fact, caused him no end of grief. Ever since its publication he had been pestered, one might even say hounded, by letters from inventors seeking to explore the subject of the electric motor, some of them hoping to achieve perpetual motion. Yet despite this flux of correspondence and the fact that he considered the electric motor in its current state of development to be little more than a "philosophical toy" like the one he himself had built, he also felt strongly that his achievement in building it had not been properly appreciated. Even so, he was inclined to deal kindly with Davenport.

"I felt considerably interested in the welfare of the Inventor," Henry wrote in his letter to Silliman, "and with friendly motives advised him to abandon the invention." Perhaps, he suggested, Davenport could exhibit it at Peale's Museum, the private hall of curiosities built in Philadelphia by the painter Charles Willson

Peale, and make a dollar or two. Other than that, Henry saw little future for the device in the world of 1835. Although he wrote a "certificat" that seemed to endorse the machine, he actually damned it with faint praise. In the "certificat" Davenport's ingenuity was again mentioned, but Henry doubted that the motor was feasible for commercial purposes. The problem, Henry pointed out to him in their conversation, was the current lack of magnetic power. In other words, the electrical output of contemporary batteries was simply too weak for an electric motor to do useful work in an economically feasible way. But at this point Davenport only heard what he wanted to hear, and he didn't want to hear that there wasn't enough electrical power in the world to make his motor a going proposition.

Continuing his pilgrimage, Davenport met and impressed Alexander Dallas Bache, Franklin's great-grandson, then a professor of chemistry at the University of Pennsylvania and later the pioneering head of the government's Coastal Survey. Bache gave Davenport a "certificat" that was less even-handed and more enthusiastic than Henry's. However, when Davenport got to Washington, neither of his certificates impressed the Patent Office. According to Davenport, his application was rejected because no one had ever patented an electrical device. This may be so, but, more important, Davenport had also failed to equip himself with a model, without which no patent could then be issued.

By the time he got back to Troy later in the summer of 1835, Davenport was famous; Eaton's newspaper article on the motor had done its work. But he was also broke. Eaton persuaded Van Rensselaer to buy the motor for $30, which got the blacksmith home to Vermont. Yet Eaton, a man who backed his bets, was not quite done with Thomas Davenport.

In September he learned that the *New York Commercial Advertiser* had written that Davenport's motor didn't work. "Totally untrue," Eaton replied. It was true that the "battery cups" had been damaged, but "I have seen [the motor] often in motion, and know it to be all that has been said of it, an astonishing application of electro-magnetism." An energetic man, Eaton took further steps. He bought all the copper wire in Troy, five pounds of it, sent it to Davenport, and announced that on October 14, the school's last examination day, when scientific luminaries would be present, the inventor would demonstrate his machine "in full motion and moved solely by electro-magnetism. It will carry three wheels weighing eleven pounds and a miniature trip hammer, as an emblem of the inventor's avocation." He wrote to Davenport, promising "a thousand spectators. I will deliver a lecture while you are working it, explaining the principles. Truth, truth is everything."[6] Davenport replied that he was out of money again, and Eaton's patience began to wear thin. "You have involved friends and it is your duty to support their pledges," he replied. No record can be found that the demonstration was ever given.

But Davenport was able to find other backers who, though vastly less eminent than Eaton, came across with a little money. Although his blacksmith business had languished, his mind was working at full tilt, and the following year he made a miniature electric train that traveled around a circular track. Using his original motor, he drilled iron and steel and turned hardwood on a lathe, and while he was waiting for something big to happen, he picked up a little extra money by giving demonstrations of his various devices. At some point he also attracted the attention of Ransom Cook, a Saratoga furniture manufacturer; in addition to making an investment, it appears that Cook made most of Davenport's subsequent motors. A model of it was finally constructed and sent to the Patent Office in 1836, together with more letters from the scientific community. The Patent Office burned down. But the next year, Davenport's application went through. He received patent number 132, and he and Cook were ready to make some real money.

Alas, the year 1837 was a terrible year for anyone to make any kind of money. The country was in the middle of its first great financial crisis. But, as always and despite the disaster, there were still a number of plausible individuals who claimed to know the secret of finding fresh funds. A New York promoter promised to raise $12,000

THE BLACKSMITH OF BRANDON   59

with a stock offering. Davenport and Cook moved down to the city and prepared to clean up, but only $1,700 of the promised cash materialized. The partners set up a workshop near Wall Street. And Davenport had a breathtaking moment of good press. The *New York Herald* reported that the Davenport motor was "A Revolution in Philosophy — dawn of a new civilization, most extraordinary discovery — probably the greatest of ancient and modern times — the greatest the world has ever seen — the greatest the world will ever see, the days of steam power, and animal power, and water power, are gone forever."

The *Herald* was wrong, and one good press day wasn't enough. Davenport attached a motor to a printing press of his own and published what was claimed to be the world's first electrical journal, *The Electro-Magnet and Mechanical Intelligencer*. It wasn't actually the first, for Sturgeon had already published one. Davenport's version does not appear to have gone into a third issue. He tried again, with a publication called *The Magnet*. It, too, disappeared without a trace. Cook went home.

There was one last, forlorn hope. An "Ohio man" came forward with $3,000 in currency issued by an Ohio bank. But President Jackson's policies, which had caused the crisis to begin with, were driving state banks out of business, and although Davenport was able to spend $10 of the money, the rest of it became worthless. Davenport, too, went home. He was never able to work again.

It can be said that Davenport failed because of a perverse miracle of bad timing. He had tried to raise money in the middle of a financial panic, and the experience broke him. But Davenport also failed for a reason that both his Forestdale neighbors and Joseph Henry tried in vain to point out. Although they didn't say it in so many words, Davenport had invented the electric motor forty-five years too soon. Henry had been right. There simply wasn't enough electricity in the world to run a practical electric motor, no matter how ingenious the inventor.

As it happened, Davenport was not alone. Across the Atlantic on December 1, 1834 — just after Davenport and Smalley had gone to earth in Smalley's Forestdale workshop — a Russian scientist named Moritz Hermann Jacobi reported to the Paris Academy of Sciences that he, too, had achieved electromagnetic rotation. Moreover, he was preparing to give his motor a practical application, or so he hoped.

In 1839, a subsidy from Tsar Nicholas in hand, Jacobi built a boat, twenty-eight feet long and seven feet wide, which he equipped with a sixteen-electromagnet electric motor attached to an arrangement of paddles. Perhaps because he could afford it, Jacobi was operating at the cutting edge of electrical science. His power source consisted of 320 Daniell cells, the powerful new batteries that were created by John Frederic Daniell in England in 1836 and used zinc and copper to generate their electricity.* Jacobi loaded a crowd of people into his craft — accounts mention "about 14" of them — set it afloat on the Neva, and achieved a meteoric speed of three miles an hour. The experiment was a failure. The weight of the batteries was too great, the speed was too low, and for the moment there was no way around the problem.

Others, on both sides of the Atlantic, were also experimenting with electromagnetic rotation. In 1838 and 1839, Robert Davidson of Scotland built an experimental electric railway car that was sixteen feet long, weighed six tons, and attained a speed of four miles an hour. In 1851 the American experimenter Charles Grafton Page received a congressional grant of $50,000 to build a full-scale train of his own, powered by one hundred of the new Grove nitric acid batteries, which incorporated platinum and dispensed with the brine. On April 20 of the following year, Page's locomotive set out from Washington to Bladensburg, Maryland, starting at a snail's pace but eventually achieving a record speed of nineteen miles an hour. Unfortunately, every time the locomotive received a small jolt, the batteries stopped working, and the problem could not be solved. Page corresponded with Davenport, although Davenport

*Three hundred twenty Daniell cells were the equivalent of a hundred six-volt alkaline storage batteries.

was flat on his back in Vermont with what appears to be a nervous breakdown, and the exchange reveals that Davenport was the superior theorist. Page still believed that practical rotation could be achieved by inertia, a dead end that Davenport had already explored. Davenport insisted that the answer was the commutator, and he was right. But neither he nor Jacobi nor anyone else ever figured a way around the battery problem. In 1835 and for many years to come, all the controlled electricity in the world came from batteries that descended from Volta's invention, and there wasn't enough tame, cheap electricity anywhere on earth to power a practical mechanical device equipped with moving parts.

Davenport and the other experimenters of his day could boost the efficiency of their machines by building larger and larger batteries, but they continually had to replace the batteries' metal cores at prohibitive expense. Moreover, as Jacobi discovered, there was the weight problem, although it did not apply to Davenport's stationary devices. Although Volta had broken through Franklin's barrier that prevented electricity from doing anything useful, no one had broken through the Volta barrier, where a trickle of power remained prohibitively expensive.

Thomas Davenport died in 1851 at the age of forty-nine. He had spent the final years of his life trying to use an electromagnet to make a piano sound like an organ.

CHAPTER FOUR

SAMUEL MORSE AND THE
INVISIBLE WORLD

Twas Franklin's hand that caught the horse,
Twas harnessed by Professor Morse.

— *popular doggerel*

I n a time of ingrained rituals and stately passages, when an ocean
voyage took weeks, as this one did, there was something slightly
unusual about one of the passengers on the sailing ship *Sully*,
outward bound from Le Havre to New York in the Year of Our Lord
1832. Samuel Finley Breese Morse, one of America's most eminent
painters — he was a founder and president of the National Academy of Design — was, in his early forties, a little long in the tooth
to be concluding the traditional grand tour of Europe's art galleries
and salons. He had just spent two years holed up in the Louvre and
traveling about Italy, studying his craft, financing his undertaking
by painting pictures for his American patrons, including the only
known landscape he ever did, and, according to his correspondence,
utterly failing to notice that France had a fully operational, up-and-
running telegraph system that stretched from Paris to the borders.

Indeed, the continent of Europe was covered with telegraph
lines. There were three thousand miles of them in France. There
were hundreds of miles in England, more hundreds of miles in Sweden, and extensive networks in Finland and the Germanies. These

were optical telegraphs that conveyed their messages visually, which meant that the manned stations had to be very close to each other. France used semaphores, their name deriving from the Greek ("sign bearer"), as did the word *telegraph* itself ("far writer"). At each French telegraph station, there were two moveable boards on two tall poles, and the changing position of the boards spelled out the message. The receiving station would then spell out the message on its own semaphores, and thus it was passed along. England and Sweden used shutters, big portholes in the station wall whose colors could be changed from black to white, spelling the message, but the pass-along system was the same. Abraham Niclas Edelcrantz, the inventor of the Swedish telegraph, had made the acquaintance of Watt, imported a dozen Boulton and Watt engines, and brought the Industrial Revolution to Scandinavia.

By the time of Morse's second visit to Europe in 1830, optical telegraphs had been around for fifty years — and they would continue to operate in France until the 1850s, when they were finally replaced by the electromagnetic telegraph. In England they were used until the 1860s, when they were supplanted by a distinctly English version of the electromagnetic telegraph invented by one of Faraday's protégés and in use nowhere else in the world. In Sweden they persisted until the 1880s, because the country's many swamps made it difficult to raise poles and string wires. But the European telegraph systems had an obvious drawback. They were useless at night and in fog and bad weather. But this, too, was being overcome by the time of Morse's second visit.

In London, Michael Faraday was at the height of his powers, building the first tiny prototypes of the generator and the transformer, making them less from theory than by trial and error, and he was bringing a promising young man along, Charles Wheatstone, who was so shy that he mumbled at the blackboard when he was supposed to be teaching a class. Faraday, a gifted speaker, delivered his lectures for him.

Charles Wheatstone had always been interested in the transmission of sound. The son of musical instrument makers, he had hung a remote-controlled lyre from the ceiling of his family's shop, and it became the sensation of the neighborhood. He invented the concertina and became known as the father of the accordion, an instrument for which the world has yet to thank him. He invented ticker tape and the stereoscope. More important from the point of view of the scientific community, he was also the first to measure, in an elegant experiment involving tiny mirrors, the speed of electricity over a wire. It was the making of his reputation.

Like Faraday and Henry, Wheatstone was not particularly interested in wealth and fame. His six-wire telegraph, which he hadn't bothered to patent, displayed its signal on five dials, and the operator used a dictionary to decipher the message. Because the needles in two of the dials had to point out a letter of the alphabet on a message board at the station, two wires were needed. Because a telegraph, to operate, has to create an electric circuit, which is closed, the third wire was needed to return the electricity to its point of origin. Three other wires made two-way communication possible. But in 1832 there was no operator and there was no message. Wheatstone had conceived his telegraph as an intellectual exercise, and there the matter might have rested had he not been approached by William Fothergill Cooke, a medical model maker and former army officer, who was extremely interested in making himself a large sum of money.

There was something slightly preposterous about William Fothergill Cooke, just as there was something slightly preposterous about Morse. After resigning his commission in the Indian Army, he set out in Germany to become a maker of anatomical models for doctors who either had no access to fresh corpses, were afflicted with short memories, or were otherwise deprived. While pursuing his anatomical studies, he was looking for some chance to make money, and in Professor Muncke's classroom at Heidelberg University he found it. To increase his pupils' understanding of electromagnetism, Muncke demonstrated one of the Russian baron Pavel Schilling's galvanometers — the deflecting needle, first observed by Ørsted, that was at the heart of a never-to-be-built Russian telegraph. On the spot, Cooke had what can only be described as a genuine Morse moment. He never explained how he got from a galvanometer to a telegraph, and the mental process is obscure and probably based in ignorance rather than knowledge, but "eureka" overtook him. Although, like Morse, he knew absolutely nothing useful about electricity, he decided to build a telegraph of his own.

Cooke, inspired by Schilling's galvanometer, settled on a design (in view of what happened later) of unusual complexity. A needle or a series of needles, deflected by electromagnetism, would be made to twitch either left or right. In Wheatstone's original conception, these left-right deflections would spell out a word in code, and the code (as Morse, too, intended) would then be deciphered by consulting a codebook. But there was another possibility, and this was Cooke's contribution: the needles could point to a letter of the alphabet, and no codebook would be necessary. Unfortunately, although he may have encountered it during the course of his studies, Cooke didn't understand Ohm's Law — and because he didn't understand it, he couldn't get his signal to pass through more than thirty feet of wire. The law had recently been formulated by an obscure German physicist named Georg Simon Ohm, and not many people in the world understood it. Simply put, Ohm's Law states that the power derived from electricity derives from the resistance of the wire down which it travels. Ohm, who died too young to witness the Electric Revolution, believed that nothing practical would ever result from his law. In fact, the Electric Revolution could not have happened if people did *not* understand Ohm's Law.

Cooke's apparatus had three dials and used five wires, we have no idea what it looked like, and for all practical purposes it didn't work — thirty feet was no distance at all. Cooke took his problem to Faraday, who could have cleared the matter up with a few sentences, but although the great scientist was impressed with the technological soundness of Cooke's transmitter and receiver, Cooke turned the conversation to perpetual motion machines before Faraday could reveal the solution to the distance problem, and Faraday remembered pressing engagements elsewhere. No closer to a solution, Cooke next managed an introduction to Wheatstone and apparently couldn't get in a word on perpetual motion before Wheatstone told him his telegraph wouldn't work. Then Wheatstone described one that would, and after considerable urging by Cooke, they entered into one of the stranger partnerships in the history of the electrical arts.

They didn't like each other, and they fought for years. Wheatstone understood the science and knew how to overcome the difficulties; soon he developed a relay that saved them the trouble of positioning a telegraph station every twenty miles along their line. A relay, as Morse later found, is the absolute key to a telegraph system. Because of Ohm's Law, the resistance in the wire will eventually make the signal undetectably faint. This was why, in their early system, the stations had to be twenty miles apart: because the receiving station, like an optical telegraph, had to copy the message down and retransmit it. Anything that cuts extra wire out of the system will lengthen the range of the signal. A relay does this by allowing a branch station to use its own battery to grab the signal and power its equipment, and Wheatstone's primitive relay answered the case.

Cooke, vastly more outgoing, was the pair's liaison to the railroads along whose right-of-way they buried their first wires and, when that didn't work very well, strung them on poles, and it was apparently Cooke who devised the message board for their new, streamlined two-needle system, the apparatus where the needles would point to the letters of the alphabet. (Wheatstone's original telegraph had six needles. They'd gotten it down to three, and two was the telegraph's final form.) It was shaped like a kite, with twenty letters of the alphabet displayed and connected by lines both horizontal and diagonal. The six needles were placed in interlocking locations along lines drawn to the letters, and when two needles pointed to a single letter, the letter was identified as part of the message. The letters c, j, q, u, x, and y were omitted. Although Wheatstone and Cooke patented their system in 1837, it won enduring fame in 1845 when it was used to apprehend a fleeing murderer only hours after he committed his crime. "Dressed like a kwaker," the Wheatstone-Cooke operator telegraphed. On the receiving end, all was bafflement, and the operator was instructed to send again. The same message appeared. It went on like that until another stock figure of the period — a bright lad — pointed out the obvious phonetic spelling; the murderer was dressed as a Quaker and, thus conveniently garbed, he was apprehended on the railroad platform as he descended from his train, to the delight of the great crime-loving British public. Morse had never heard of Wheatstone and Cooke. And they had never heard of him.

He was a dissatisfied man, Samuel Morse, and dissatisfaction was a distinctly American trait. He had made the grand tour once before, as a youth just graduated from Yale. Then he had confined himself to England, mostly to London, where he had apprenticed in the studio of Benjamin West, the American who served as the official royal artist. Morse's parents had provided the money, and it ran out — too soon, in Morse's somewhat embittered view. In American eyes, Europe was the source of all wisdom, knowledge, and culture. European training was essential for the artist's calling, and Morse didn't think he'd gotten enough of it. Indeed, he seemed to think he hadn't become European enough. His fellow Americans didn't think much of American artists, and although he was one of the most prominent of them, he'd been reduced to painting portraits rather than the historical scenes that were the highest and truest examples of his art. Americans didn't buy historical pictures unless they were European. In fact, Americans didn't buy much art at all. When they paid big money for it, it was in the form of a ticket for a gallery where a large painting of a compelling subject — Hell, for example — was on view.

This was the reason for Morse's second trip across the ocean. Nagged by the notion that there was something wrong with his paintings — why else wouldn't his countrymen buy them? — he had gone to Europe to perfect his craft. Perhaps, too, he had a feeling that some profitable European attributes would rub off on him, which may have motivated him to blaze new trails in France and

Italy, the homes of high art, rather than confine himself again to England, a country many Americans still called "home." All his life Morse had wanted to make a great deal of money. This, too, was an American characteristic. Europeans often commented on it, the English usually despised it, and even now, as the *Sully* plowed the waves westward toward New York, the Frenchman Alexis de Tocqueville was touring the United States, gathering material on the very American trait of acquisitiveness, one of the dominant national passions. In his luggage Morse had a large unfinished painting that he planned to exhibit. It would, he hoped, be the making of his fortune.

His father, a friend of the dictionary author Noah Webster, was the principal Congregational clergyman in Charlestown, Massachusetts, and the author of the most popular geography text in the United States, where he had written that America had no tall mountains. When not distracted by the growing Unitarian heresy, which eventually cost him his pulpit, the elder Morse and his wife were deeply involved in the moral instruction of their children. Finley (as his family called him), born in 1791, was a father's despair. At Phillips Academy in Andover, the lad did well when he applied himself, but his tutor was compelled to write that Finley was "regardless of truth — idle — and at the bottom of [his] class." Nor did he thrive at Yale, a college chosen because of its religious orthodoxy, where there was no school of fine arts and the only classes that seemed of interest to him were those that dealt with electricity. From the professor of natural philosophy, Jeremiah Day, young Morse learned a passage from Enfield's *History of Philosophy*: "If the [electrical] circuit be interrupted, the fluid will become visible, and when it passes it will leave an impression upon any intermediate body." Electricity had remarkably little to do with the life he chose to lead until he reached his forty-first year. But he was a close, if selective, observer of his surroundings, and one of the things he observed was the pernicious effect of time on human communication. For example, he was in London when the British government caved in to American demands that British warships stop kidnapping American seamen, his hosts made certain that he was aware of the fact, the word did not reach America in time, and the Americans plunged into the War of 1812.

Following immediately on his graduation from Yale, Morse's art student years in London between 1811 and 1815 saw great ferment in England, as did his second trip, when he added France and Italy to his itinerary. The Industrial Revolution was going full blast, and England virtually owned it all. So too, although on not quite so exclusive a basis, did England seem to own the Electric Revolution as personified by Michael Faraday, though both revolutions were beginning to display symptoms of the serious slowdown, verging on stagnation, that would afflict the country's industry by the end of the century. On the industrial front, the British were committing a form of entrepreneurial suicide as the industrialists emulated their aristocratic betters and retired to country estates with fortunes large enough to allow their descendents to live in rural idleness. The ownership of land and the possession of ample leisure time were the signs of social status, and the ownership of an industrial empire wasn't. Meanwhile, the study of electricity never fully emerged from think tanks like the Royal Institution, and as a result, the British blew the largest lead in scientific history. The intellectual heir of Michael Faraday was the Scot James Clerk Maxwell, who in 1873 laid the mathematical foundations for electrical theory that survive to this day. In America, the heirs of Joseph Henry were Morse and all who came after him: scientifically unlettered men who exploited the theories and discoveries of others to create useful devices and make their fortunes. Faraday and Henry gave their findings to the world. Morse and his successors were out to make a buck.

* * *

On Morse's first trip to England as a very young man, the engines of Watt were proliferating over the landscape, Humphry Davy was lecturing at the Royal Institution, and Morse left little sign either in his letters home or in his lifelong verbal recollections of his London student years that he noticed either of these things. But he was an American in a foreign country that soon became an enemy nation, and it could not have escaped his attention that Andrew Jackson

defeated the British invasion force with great slaughter at New Orleans after hostilities had officially ceased. Still, the Morse of 1811
through 1815 was a young man of his time. News had always traveled slowly, there was nothing unusual about that. And he had
other, far more compelling matters on his mind.

With another art student, he took rooms in the West End of
London, and almost immediately upon taking up his studies in 1811,
young Morse detected unmistakable signs that he possessed not only
an ambition but a genuine talent. Benjamin West, the American
who was King George's personal artist, had consented to give him
lessons. Soon West allowed him to work on the immense historical
canvases he was painting for his royal patron, although not the important parts, to be sure. Morse painted the manes of horses. But it
was a sign of favor. And there were others as well. Whenever Morse
brought West one of his own pictures, West invariably said, "Go and
finish it." Often, when Morse protested that he *had* finished it, West
would repeat himself. Indeed, the simple, baffling directive was just
about the entire sum of West's instruction. But West didn't throw
him out, and he didn't advise him — in a kindly manner, of
course — to look for another job. Morse sensed, and soon he had evidence when West encouraged him to branch out into sculpture
while plugging away at his painting, that he had won the court
painter's favor. Other artists, more established than he, seemed to
enjoy his company, and they said they enjoyed his work. His student
days were far from over, he felt, and he sensed he still had much to
learn; for years thereafter, until he returned to Europe in 1830, he
complained to friends that he hadn't been given enough time during his apprenticeship. He was enjoying himself — he never enjoyed himself more — and he was attracting attention. But his
parents were supporting him, their resources were at an end, and
the time had come for him to return to America where, he soon discovered, the life of a serious artist in the European sense was a life
of almost unremitting toil and poverty. Just before he left in 1815,
he heard guns in Regent's Park saluting the victory at Waterloo.

In England his *Dying Hercule* — a sculptural rendition of a student painting — had won a considerable prize, but back in America, he found, there was absolutely no market for the historical paintings that Morse and his former mentors regarded as the highest expression of the painterly art. Americans, he discovered to his enduring horror, were vastly more interested in themselves than they were in serious art. Morse, trapped in America by his lack of money, tried to adjust himself to his countrymen's tastes. But in his art he never seemed to hit upon the correct formula.

Morse settled in Count Rumford's hometown of Concord, New Hampshire, where he married the local belle, Lucretia Pickering Walker, a Count Rumford relative, and later moved to Charleston, South Carolina, where he had relatives. Morse was able to make a respectable living from portraiture, a popular and lucrative form of work he regarded as pedestrian, although he was rather good at it. Then Charleston dried up as a source of commissions when his portraits, though good ones, went out of fashion and a fickle public no longer visited his studio. He toyed with becoming an architect but gave up the idea; he considered joining the Episcopal priesthood and decided against it. While he was in Washington, D.C., painting his memorable portrait of Lafayette during the old hero's farewell visit to America, his young wife suddenly died. By the time he received the news and returned to New Haven, where his family had moved, she had been buried. He had never been able to provide her with a home — they lived with his parents, now in theological exile from Massachusetts, where Unitarianism had triumphed despite his father's many thunderbolts. He was unable to provide for his three children. He placed them in the care of an old family retainer and later farmed them out to his relatives. He moved to New York, where he expected to immediately become the premier artist of the city. He did not. He continued to paint the portraits he considered beneath him, and did them well, making a modest living, although not the one he had once imagined for a serious artist practicing serious art.

Although he had found that Americans did not commission historical paintings for their personal collections, they would pay to see them in a rented gallery, and he tried his hand at it. His first such painting, *Congress Hall*, a huge one and arguably the best he ever did, depicted the House of Representatives at twilight, with the legislators — among them Philip Van Rensselaer — deployed at their desks as the clerks are lighting the great chandelier. In the visitors' gallery, he placed a Pawnee chief who happened to be in town for a treaty negotiation, his father, and Professor Silliman.

In New York he rented space, set up his picture, placed advertisements in the papers, worked up a key so the customers could identify their elected representatives, and waited for the money to roll in. It didn't. The mass of New Yorkers weren't overly interested in paying to view their elected representatives; they much preferred to plunk down their money for a look at Rembrandt Peale's *Court of Death*, and Morse's *Congress Hall* ended its run at a loss. But Morse was undeterred. People would, demonstrably, pay good money to look at huge paintings; he had only to find an appropriate subject. On his second trip to Europe he believed he had hit upon it. In Paris he had renewed his acquaintance with the Revolutionary War hero Lafayette, a former subject whose Morse portrait hung in New York's City Hall. Lafayette implored him to delay his return to America; he would have more congenial company on a later boat. But Morse stuck to his plan and sailed on the *Sully*. With him he had a giant painting of the gallery of the Louvre with his friend Fenimore Cooper prominent in the composition and the gallery's famous collection — actually, the paintings Morse *believed* should have been in the gallery's collection, for he was not fond of French art — reproduced and scattered about the canvas. Then, on the *Sully* and by the merest chance, he found himself gobsmacked with an absolutely wonderful idea for making his fame and fortune, although it had nothing to do with art. He experienced a eureka moment.

The eureka moment — the blinding flash of inspiration that leads directly to the invention of a world-changing device — occurs far less often than the folklore suggests, but sometimes it happens. Davenport experienced one such moment when he saw Henry's magnet at work. Samuel Finley Breese Morse seems to have had his own while he was eating lunch. Although Morse shamelessly fictionalized much of his life as it was lived after this seminal moment, the event actually occurred. For once — a rare occurrence in Morse's subsequent career of stealing credit from his associates or ignoring their contributions — all the witnesses but one agreed with his version of the story.

What happened was this: One day, over lunch on the ship, the talk turned to Ampère's recent discoveries in electricity, and a question was put to the table by a certain Mr. Fisher, a lawyer from Philadelphia. Was not the flow of the electricity, Fisher asked, retarded by the length of the wire?

In 1832 it was a good question. Indeed, certain promising experiments had come to a dead halt precisely because electricity on a wire appeared to peter out after a few feet. In order to understand why this was so, it was important to understand Ohm's Law. By a coincidence that is little short of astonishing, one of the few people who did understand it happened to be seated at the luncheon table of the *Sully*.

He was a young medical doctor named Charles Thomas Jackson, returning from a course of study in his profession at the Sorbonne and in geology and mineralogy at the Ecole des Mines. He was a brilliant man but a strange one. In a number of his surviving photographs, he seems distinctly peeved, an expression that did not bode well for his enemies, of whom he soon had many. But he was good company when he chose to be, and on the *Sully* he appears to have been both agreeable and informative. Addressing the subject, he informed his fellow travelers that electricity passed instantly over any known length of wire; Franklin, for one, had demonstrated it. Sparing his audience, Dr. Jackson did not explain Ohm's Law. If he had, he might have saved Morse years of grief. But Morse was instantly excited by his remarks.

"If this be so," Morse recalled himself as saying, "and the presence of electricity can be made visible in any desired part of the circuit, I see no reason why intelligence might not be instantaneously transmitted by electricity to any distance."[1]

Morse was instantly seized with an inventive fervor, and he was not exactly a stranger to it. As part of a long, doomed series of attempts to make the fortune that had ever eluded him, he had tried to create a machine for cutting sculptural marble. So had a number of other people: Watt, for one, had spent the twilight of his life pottering with just such a device. But he had failed, and so had everyone else who bent himself to the task. Another invention, which Morse co-patented with his brother Sidney, was a fire-fighting apparatus jokingly called Morse's Patent Metallic Double-Headed Ocean-Drinker and Deluge-Spouter Pump-Box. It failed on its first trial. He knew a little about electricity and electromagnets, not much, but some. He had briefly studied the subject at Yale, and his neighbor and friend when he later lived in New Haven with his wife and parents was Benjamin Silliman; on at least one occasion they had taken a battery apart and studied it. It therefore seems almost impossible, but it also appears to be entirely true, that one of America's foremost artists, a somewhat otherworldly man pushing middle age, who had never made a successful invention in his life, should have risen from the lunch table on the *Sully* with an idea for an electric telegraph dancing in his head. He quickly sketched a possible device in his notebook — the device, in fact, he would eventually build.

He believed that no one in the world had ever thought of such a thing, and the notebooks suggest that he foresaw little difficulty in assembling an apparatus and making it go. It would require an electromagnet. This was his first breakthrough, and an original one. But although no one else had thought of using an electromagnet, Morse was mistaken in his belief that he alone had conceived of the electric telegraph. More than a decade earlier, the Russian nobleman Pavel Lvovich Schilling von Cannstadt became impressed by the German scientist S. T. von Sömmering's 1809 telegraph of twenty-six wires, joined the inventor in a collaboration, and by 1820 had produced a system that picked out its messages with a single moving needle. It was the first electromagnetic telegraph, and it came tantalizingly close to commercial deployment. Schilling obtained an audience with the tsar in 1836, the tsar was deeply impressed and gave orders to construct a network, but Schilling died, and his project died with him.

Schilling had stumbled on the idea of an electromagnetic telegraph, but there was also nothing new about the idea of an electrical version of one. The possibility of electrical transmission had been obvious ever since Franklin strung a line over the Schuylkill River in Philadelphia and used it to ignite alcohol fires on both banks, to the delight of his assembled guests. Indeed, more than sixty previous electrical telegraphs had been conceived at various times, in various places, by various individuals, some of them so obscure that their first names, together with anything more than a vague description of their devices, seem to have been lost. This deplorable situation also seems to have prevailed in 1832. In any event, telegraph work was not exactly a hot topic at the average family hearthside, especially if the head of the family was a professional artist. Unknown to Morse, who all his life clung to the belief that he, and he alone, had conceived of the electrical telegraph, someone named Reizen may have constructed a friction-electric telegraph in 1794 that used seventy-six wires, Agustín de Betancourt built an experimental line between Madrid and Aranjuez in 1798, and another Spaniard named D. F. Silva had a twenty-six-mile line powered by Leyden jars. Morse may have taken an original course in planning his electric telegraph, but he was far from being the first to conceive of one. In the 1750s a Scottish magazine published a letter, cryptically signed "C.W.," that described a version of an electrical telegraph system right down to the poles that carried the wires and the glass insulators that kept them from shorting out. But neither the telegraph described in 1752 nor its predecessors had any widespread practical use, because Leyden jars and friction machines were the only available power source; Volta did not invent his battery for another forty-four years. Claude Chappe, the inventor of the optical telegraph popular in France, had considered using electricity on a line in the 1790s and had a good notion of a possible system, then decided he didn't have an adequate power source and settled on the mechanical semaphore — in effect, a moving board fastened to an upright pole — as a superior solution. In 1816 Sir Francis Ronalds, belatedly knighted for his achievement, built a static-electricity-and-pith-ball telegraph in his Hammersmith garden near London. The electricity made the pith balls move, and the pattern made by the balls spelled out the message in code. Ronalds offered his invention to the Admiralty, was flatly refused, and turned to the study of the weather. In America, Harrison Gray Dyar experimented at a Long Island racecourse — it enabled him to lay out his wires — with an electrical telegraph that chemically recorded his message on litmus paper, but ran afoul of the law when it was suspected that he was "sending secret communications in advance of the mail." Dyar abandoned his planned line between New York and Philadelphia, moved to France, and made a fortune with nonelectrical inventions.

By 1830 there was at least one short electrical telegraph line in the fragmented Germanies at Göttingen University. It ran between the observatory and the physics lab, and great things would soon begin to happen there. And in England, Faraday's protégé Charles Wheatstone had conceived of a six-wire system. Morse knew none of this. Most people didn't.

After Morse drew a picture of his proposed device, the next challenge he confronted on the *Sully* was the question of just how the projected telegraph would communicate over the wire. As the ship crept across the broad Atlantic, he buckled down to the task and devised a binary code of dots and dashes. It was his second great breakthrough; no one else had conceived of a binary code. He devised a language: the number one would be a single dot, five would be five dots, and six through ten (or zero) would be the same dots followed by a dash. Then he went swiveling off in the same direction pursued by many of the other code makers who had dreamed of the electrical telegraph. Combinations of numbers would stand for words. Fifty-six, for example, would mean Holland. Fifteen would be Belgium. To decipher a message, a codebook would be needed, and he did not consider just how thick that codebook might have to be, or how long a hapless telegraph operator might have to spend in winkling out a message.

For the rest of the weeks-long voyage, he was all afire with his proposed invention. On debarking, he congratulated the captain on the latter's ship, where, he assured that nautical gentleman, the electrical telegraph had been conceived. History does not record the captain's reaction, but Morse, meeting his brothers on the dock, could speak of little else. He speckled the hearthrug of his sister-in-law's house with drops of molten lead as he cast a component for his new invention. Then, for the longest time, he set it aside. Samuel Finley Breese Morse always had a hard time sticking to something.

Back in New York once more, he again rented a showroom, printed up tickets, took out advertisements, and waited for the money to roll in. But Americans proved no more interested in looking at his *Gallery of the Louvre* than they were in contemplating *Congress Hall*. His destiny continued to elude him.

By now, in the mid-1830s, after two trips to Europe and an itinerant career in various parts of America, he was the professor of sculpture and painting — the first in the country — at the new University of the City of New York, later renamed NYU, where he earned his keep by collecting fees from his students. He had a steady income that, in typical Morse fashion, he managed erratically. He had rooms of his own — "a perfect shower bath," he complained, and the chimney drew poorly; it was an altogether wretched place — in the tower of the university's new building on Washington Square. But he had leisure, and he returned to the idea he had first evolved aboard the *Sully* — the telegraph. Now, he decided, it was time to buckle down to work on it. But first, he decided to run for mayor.

* * *

By all surviving accounts he was a kindly man, although his upbringing among the always contentious Protestants of New England had given him a sometimes unfortunate taste for disputation. Still, he had ample social graces, and they had won him many friends. His parents had been dedicated Federalists, but Morse, considering

himself more enlightened, styled himself a Jeffersonian Democrat.
His contemporaries, however, regarded him as a conservative.

His roots were deep in an Anglo-Saxon, Huguenot, and Dutch
America that was now passing away, and his experiences among the
charming and cultivated planters of Charleston had done little to
enlighten him. He thought slavery was an excellent idea, and he
was deeply suspicious of the new Irish immigrants who had begun to
arrive on his shores. They were disturbingly poor and, worse, they
bore with them the fatal contagion of Catholicism. Morse was not
only a son of New England but the son of a New England divine
who had seen a dark threat in the largely imaginary machinations of
the Bavarian Illuminati. In Europe, especially in Italy, he had seen
the Church of Rome as the instrument of autocratic government.
Now, with the arrival of the Irish, who were proving alarmingly
adept at the art of politics, the seed of Rome had been planted in
America. And Morse could never resist a good fight.

To say that his politics were seriously misguided is both correct
and incorrect. In 1836, when he first stood for office, Native Amer-
icanism was a perfectly respectable, ostensibly nonpartisan political
movement, and it was a powerful one with many influential adher-
ents. The Native Americanists were deeply anti-Catholic, which
meant they were deeply anti-Irish and therefore against Irish immigration. Although history proved them wrong within Morse's own
long lifetime, he never saw any reason to alter his views on either
the Papacy or the South's peculiar institution. During the Civil
War, an ageing man, he was a prominent Copperhead, a Northern
sympathizer of the Confederacy and its cause.

He produced a pamphlet called *Foreign Conspiracy against the
Liberties of the United States* and was credited with inspiring a riot by
Native Americanists in his hometown of Charlestown, Massachusetts, when the Reverend Lyman Beecher delivered the incendiary
sermon "The Devil and the Pope of Rome." Morse wrote articles
and more pamphlets, revealing that he thought the burning of a
Baltimore convent — possibly the greatest outrage ever committed
by the Nativists — had been a wonderful idea. When Philip Hone,

a former New York mayor known to posterity as a celebrated diarist, passed on the chance to become the Nativist standard bearer, Morse was the obvious choice. But he wasn't much of a campaigner, and it wasn't the Nativists' year. They did eventually win the mayoralty, but not on Morse's watch. He got 1,500 votes, and he lost by more than 33,000. Curiously, this did not diminish his appetite for either Nativism or politics. It did, however, leave him again with time on his hands. He had been working on his telegraph, not steadily, and now he had a working model. Unfortunately, it didn't work very well.

It may have been Robert G. Rankin, a lawyer, who first got a glimpse of the electromagnetic telegraph. Passing along the east side of Washington Square in the autumn of 1835, he heard his name called, recognized Professor Morse, and received a rather puzzling greeting.

"I wish you to go up to my sanctum," Morse said, "and examine a piece of mechanism which, if you may not believe in, you, at least, will not laugh at."[2]

Morse's rooms were on the third floor. Entering, Rankin saw what appeared to be a melodeon, a keyboard musical instrument, surrounded by coils of wire. Morse explained its use: When the keys were depressed, they sent electrical impulses over a wire in the form of dots and dashes. A long pulse of electricity was a dash, a short one was a dot, and it was as simple as that. At the end of the wire was an electromagnetic receiver. The electromagnet, turning off and on, worked a device that recorded the dots and dashes on a piece of paper. Unfortunately, we have to take Rankin's word for it. Morse apparently never committed the likeness of his first actual telegraph to paper. Now, he awaited his friend's reaction.

"Well, professor," said Rankin after a moment of reflective silence, "you have a pretty play! theoretically true, but practically useful only as a mantel ornament."

It wasn't the idea of a telegraph that stopped the thinking processes of a cultivated and intelligent man like Rankin. It was the electricity. In 1835 almost nobody knew the faintest useful thing about electricity. But telegraphs themselves were old hat, and every cultivated and intelligent man like Rankin knew what they looked like. The Paris that Morse had visited six years previously was the center of the largest and most efficient telegraph system in the world. There were telegraphs in America, too, where a line linked the New York and Philadelphia stock exchanges. Financial news passing between the two cities reached its destination in three minutes. In the older cities on both American coasts, whenever you stumbled upon a Telegraph Hill or a Telegraph Avenue, that was where they were. Well before the nation's legislators ever heard the name of Morse, Congress was planning to run one from New York to New Orleans. A telegraph bore no resemblance to the device — whatever it was — in Morse's rooms.

When Cooke and Wheatstone received their British patent in 1837, Morse was still an object of pity. It was sad, his colleagues said, that he hadn't received a commission to paint a mural in the national capitol building; indeed, one of the artists chosen for the task, Henry Inman, attempted to resign in Morse's favor. Morse was, after all and as he had foreseen, one of the foremost painters in the country, one of the few who were classically trained and a proven master of the historical genre. On those merits alone he deserved the job, and he badly needed the money. But Inman's resignation was declined, and as a consolation Morse's friends raised a subscription of $3,000 to pay Morse for an historical painting — as the subject, he settled on the signing of the Mayflower Compact — that would *not* hang in the Capitol. It was unfortunate, most right-thinking men of the Protestant persuasion agreed, that he had been drubbed so soundly in the mayoralty race; he was, after all, a decent fellow, and his heart was in the right place where the Irish and Catholic Questions were concerned. And it was simply bizarre, the NYU community agreed, that a man of his talent should waste his time — and neglect the high calling of his art — to tinker with some silly telegraph that no one understood very well. But at NYU, it turned out, there was also at least one person who understood the silly device far better than Morse.

* * *

Like a number of American scientists of the period, Leonard Gale had started as a medical doctor. Now he was a professor of geology and mineralogy, but his talents did not end there. In January 1836, when Gale first saw Morse's telegraph, he knew at once what was wrong.

At some point, Morse had gone back to his *Sully* notebooks and revised his machinery again. In place of the melodeon-like keys, the transmitter was now based on his original idea of a portrule, a printer's device resembling a flat music box with a crank on one side that was probably obtained from his brother Sidney, who was a publisher of the *New York Observer*. Pieces of sawtoothed, cast iron type were placed on a moveable arm at the top, each bit of type representing a configuration of dots and dashes. When Morse operated the crank on the side, the sawteeth lowered and raised wires in cups of mercury, completing an electrical circuit that sent an impulse to an electromagnetic receiver mounted on an old canvas stretcher. The message was recorded on a strip of paper drawn through the receiver by clockwork. It was not an elegant device. Indeed, it was so nondescript that it was almost impossible to describe. And Morse couldn't get his wire to transmit for more than about thirty feet.

Gale knew why. Morse, like Cooke, didn't understand Ohm's Law. Now, whether or not Gale himself understood it is an interesting point but also a moot one, because he understood the scientific writings of Joseph Henry, who did. Moreover, he knew the man, although not intimately; in 1833 he had helped Henry experiment with a battery on the corpse of an executed murderer in Morristown, New Jersey. And what Henry knew and Gale understood was this:

Contrary to common sense, it isn't the amount of current pumped down a line that determines the amount of work that can be done or the distance at which it can be performed. It is, instead, the *way* in which the electricity is produced, and if the correct way is chosen, it is possible to send a weak but usable current a very long way indeed. An array of batteries — several of them, not just one — would overcome the resistance in the wire that Ohm had discovered and incorporated in his law, the resistance that stopped both Cooke's and Morse's signals after they had traveled thirty feet.

Morse, following common sense, was using a single large battery with the usual two poles, positive and negative. Gale persuaded him to change to a number of smaller batteries with multiple poles. No one quite knew how they worked, but they did. And, Gale observed, Morse was using the wrong kind of electromagnet.

Morse had first encountered the electromagnet when he attended the lectures on Sturgeon's new device that were delivered at the New York Athenaeum in the early months of 1827. It was at these same lectures that Henry first became acquainted with the magnet and began his discoveries. But Morse and Henry did not know each other, and Morse had learned nothing about electromagnets since then. His 1836 magnet was a loosely wound Sturgeon magnet and utterly unsuitable for his purpose, because it was neither powerful nor sophisticated enough. Gale taught him how to build a tightly wound Henry magnet and urgently called his attention to Henry's seminal 1831 paper in *Silliman's Journal*, where Henry laid out his recent discoveries about the electromagnet. There, Gale said, was everything Morse needed to know.

In later years Morse repeatedly claimed that he had never read Henry's paper. A judge, carefully examining Morse's telegraph apparatus and then pondering his denial, flatly refused to believe him. The whole system was based on Henry's battery array and Henry's magnets. Morse's magnets were Henry's magnets, they were set up in the way Henry had described, and when confronted by Henry, Gale — but not Morse — had no choice but admit it or resign from the scientific community.

It had taken Cooke and Wheatstone mere months to build their telegraph and get it into operation; their initial problem was not the technology but finding a right-of-way. It had taken the impoverished and distractible Morse nearly five years from the inspiration in the dining saloon of the *Sully* to the fortunate arrival of Leonard Gale, and he still didn't have a commercial product; he had an electromagnet mounted on a canvas stretcher, a bit of clockwork, a portrule, a hand crank, and some mercury cups. But with the arrival of Gale, focus finally began to creep into his life. With Gale he became fortunate in the partners he began to assemble to carry the telegraph project forward.

Once Gale's improvements were incorporated, Morse's telegraph could send a signal for a thousand feet. Then reels of wire were set up in Gale's lecture hall, and they sent a signal for ten miles. The thing was crude, but it worked, and it was high time. Morse was being overtaken by Wheatstone, Cooke, and their own complicated device. At some point he received word of their telegraph through the scientific journals, and he may have heard word of the English apothecary named Edward Davy who had challenged the Cooke and Wheatstone patent on the grounds that it was similar to his own. And in the Germanies, Karl August von Steinheil, a Munich astronomer, was making great, groundbreaking strides with the prototype electric telegraph at Göttingen. To make matters worse, the same financial panic that was the ruin of Davenport's hopes caused Morse's painting clients to find other uses for their funds, which gave him more time to work on his telegraph but less money for food. Understandably, Morse felt a certain urgency. And then, while demonstrating his telegraph for an English visitor to the university, he found his salvation. It came in the form of a student, not his own, named Alfred Vail.

* * *

Like the young Joseph Henry, Alfred Vail was old for a student. He was thirty and the scion of the New Jersey family that owned the Speedwell Ironworks, where the engine for the *Savannah*, the first transatlantic steamship, had just been built. New Jersey was big iron country in those days, his family were big iron people, but Alfred Vail had little desire to follow in their footsteps. He had aspired to the ministry, but his health had broken down and now he was a student at NYU, trying to make something of his life. Looking in on Morse's demonstration, he suddenly discovered his destiny.

"I returned to my boarding-house, locked the door of my room, and gave myself up to reflection upon the mighty results which were certain to follow the introduction of this new agent in meeting and serving the wants of the world," Vail wrote. "And upon this I decided in my own mind to SINK OR SWIM WITH IT."*

He approached Morse and offered money and help. His father, Judge Stephen Vail, would put up $2,000, although it was less an investment than a bribe to lure his son back to the family business. The younger Vail himself was a talented mechanic, and he would help Morse with the actual making of his device; it was clear that Morse needed all the help he could get. And Vail did more. His letters, on file at the Historic Speedwell site in New Jersey, provide an invaluable portrait of the actual making of the Morse telegraph and of the difficult character of Morse himself.

A deal with Morse was soon struck and put into writing, although it's hard to understand why the shrewd Judge Vail signed such a thing unless, as is probable, he was simply trying to open contact with his son and saw no particular future in a middle-aged artist who proposed to talk with lightning. Although it was Gale who had solved the resistance problem, and the Vail family who put up the money, Morse took the majority position in the new partnership. The contract stipulated that all inventions, no matter who had made them, would be patented in Morse's name. Morse remained in New York, where he soon had so little cash that he didn't eat for twenty-four hours and was reduced to borrowing ten dollars from one of his students. Vail, having hitched his wagon to the Morse star, returned home to Morristown, where he and a shop boy, William Baxter, proceeded to invent the devices that would turn Henry's unpatented electromagnets into a working telegraph.

While Morse labored over a dictionary that would translate the elaborate code he was devising, Vail discarded the canvas stretcher and built fitted boxes for the telegraph components. He also found Morse's printing apparatus unsatisfactory, tried pencils, then pens,

*Unless otherwise noted, all Alfred Vail quotations derive from the Historic Speedwell Web site.

then a metal-tipped embosser, and finally decided on no printer at all, instead settling on an annunciator that clicked off the dashes and dots and could be read by ear. Dissatisfied too with Morse's portrule-transmitter, he invented the telegraph key, a little metal catapult with a metal fingerpad at one end and a pointed, downward-facing electrical contact on the other, that would tap out the message. (Out of deference to Morse, who hated the key, he still used the portrule as a transmitter when Morse was present.) The telegraph had now reached the form it would retain for the next 150 years. Only one thing remained to be added. According to a well-supported family legend, Vail also visited a local printer's shop, examined how the trays of type were arranged, and learned the order of frequency of the alphabet in English. E was the most commonly used letter. Vail assigned it a dot and went on from there. Today, a preponderance of opinion identifies Vail, and not Morse, as the author of the Morse code. Certainly the code Morse was inventing across the river from his New Jersey partners bore no relation to it. He was still revising his dictionary years later, long after the code that bears his name had become the standard of the world.

"Professor Morse was very much inclined to insist on the superiority of his own plans and methods — if for no other reason; because they were his," the shop boy Baxter later said. "[But] as we became acquainted with Morse it became evident to us that his mechanical knowledge and skill were limited, and his ideas in matters relating to construction of little value."[3]

He was, however, extremely proficient at taking revenge.

* * *

While Vail was busily inventing the code and improving the telegraph, he and Morse were separated for the better part of a year. But as January 1, 1837, approached, a time of decision neared and an act of kindness was prepared. Aware that Morse was in serious financial trouble, Judge Vail commissioned portraits of himself and his wife as a way of providing the improvident professor with a little money. But January 1 was also the day the improved telegraph mechanism was scheduled for completion. Judge Vail had committed $2,000, in the 1830s a considerable sum, to the project at a time of economic distress, the neighbors had turned to mockery and spoke of a mad "lightning machine," and if the judge was inclined to treat Morse kindly, he did not gaze with unmixed kindness on his errant son. Judge Vail believed that his son suffered from a serious character defect: in all his life, Alfred had never been able to finish anything. As time passed and no completed telegraph appeared, the judge's temper grew short and his countenance darkened. Alfred and Baxter began to avoid his company, making spies of Alfred's sisters so that the judge could be avoided at dinnertime. The Vails' domestic situation grew increasingly grim, and Vail had still not produced the promised telegraph. And Morse also had to be dealt with.

"The artist realized his cumbersome picture-frame equipment was to be replaced by the practical and simple Vail instrument," George P. Oslin wrote in his definitive 1992 study *The Story of Telecommunications*. "Morse became so upset, Baxter said, that he became ill and was in bed for some weeks at the Vail home."

The Vail apparatus was beautifully crafted, and it bore no outward resemblance to the telegraph Morse believed he had invented. Vail and Baxter had noticed a pattern in Morse's behavior: every time he saw one of Vail's improvements, he seemed to get sick.

But at last, only a few nerve-wracking days overdue, the telegraph was ready. Baxter was so excited that he ran into the cold weather with no coat. He found the judge in a brown study, dressed to go out but staring grimly into space. At Baxter's summons he immediately roused himself and came to Alfred's workshop. Morse, too, miraculously rose from his sickbed, and the test was prepared. The judge was asked to compose the first message. Leaning toward his son's ear, he whispered (so Morse, the recipient, could not hear), "A patient waiter is no loser." Alfred loaded the portrule and fed it into the transmitter. Morse then read out the tape at the receiver, and the judge appeared to experience a moment of emotion. It probably wasn't exclusively about the telegraph. His son, perhaps halfway through his life given the grim statistics of the times, had taken on a task, and he had completed it.

There would have to be a demonstration for the skeptical townspeople — the judge's reputation demanded it — but first a quantity of copper wire was required, and none of the usual sort could be found. Eventually hat wire, used to support the ladies' popular "skyscraper bonnets," was pressed into service, and the town's supply was quickly exhausted. Miles of hat wire were threaded through the exhibition hall, and people came from far and wide to witness the test, many of them hoping to see the son of the stern and successful judge fall on his face in public. But the test was a success, although not exactly the mystical moment depicted in later popular accounts. Morse was always very much the hero of his own story, but on this occasion his showmanship deserted him. The first public message of the telegraph was not, as legend sometimes has it, "Attention the Universe! By kingdoms, right wheel." That came more than a year later, and it was a joke perpetrated on a member of another telegraph audience by some of his friends. Instead, according to the local paper, the *Morristown Jerseyman*, the first public message of the Morse telegraph was "Railroad cars just arrived, 345 passengers."

Other demonstrations followed, accompanied by Vail's growing doubts about Morse, especially regarding Morse's freehanded use of the Vail family's money. The professor, Vail wrote to his brother George, a silent partner in the venture, "is altogether inclined to operate in his own name, so much so that he has printed 500 blank invitations in his own name, at your expense." A later generation would say that Vail was beginning to get Morse's number. Moreover, Morse began to refer to Vail as his "assistant." There may have been more than a little irony in the message when Vail wrote, "I feel, Professor Morse, that if I am ever worth anything it will be wholly attributable to your kindness." If so, Morse did not detect it. And there was, of course, the signed agreement, which stated that unless Morse received sole credit for any and all improvements to his electromagnetic telegraph, all bets — and all hopes of recovering the Vail family's investment — were off.

* * *

There was, of course, a problem, and it was not small. While Morse had pottered around, gotten himself stymied by Ohm's Law, and taken time out to run for mayor, the Europeans had gained a five-year lead in developing their own telegraphs. Indeed, the Wheatstone and Cooke needle telegraph was already up and running on an experimental basis. "The celebrated Gauss," the Franklin Institute reported to the secretary of the treasury, who was charged with seeking out a possible telegraph system for the government, "has a telegraph in actual operation, for communicating signals between the University of Göttingen and his magnetic observatory in the vicinity." The institute was apparently unaware that Karl August von Steinheil of Munich was not only improving the Göttingen device but was experimenting with sending telegraph messages through a railroad line. "Mr. Wheatstone, of London, has been for some time also engaged in experiments on an electrical telegraph." To an informed American scientist, Morse's telegraph was no new thing, but the scientists at the Franklin Institute in Philadelphia swiftly put their finger on the feature that would give Morse's device a decisive advantage. "The plan of Professor Morse, is, so far as the committee is informed," they wrote, "entirely different from any of those devised by other individuals, all of which act by giving different *directions* to a magnetic needle."[4] A needle telegraph like Wheatstone's required a great deal of work: manipulating the dials, reading the dials, transcribing the message. Morse's telegraph, by contrast, was simple to use, especially after the invention of Vail's key and annunciator, and Morse's telegraph automatically recorded the message. But the European systems were up and running, and the Morse telegraph was not.

By a happy chance, while the Congress of 1837 was no more sophisticated in technological matters than any subsequent Congress, it looked favorably on the idea of an American telegraph. The telegraph the legislators had in mind was a French semaphore, an English-Swedish shutter system, or something else that was looked at rather than listened to. Still, Morse had in hand the endorsement of the Franklin Institute, the nation's capitol (with the exception of

Congressman Millard Fillmore) was greatly taken with his machine, and the chairman of the House Commerce Committee, the thirty-two-year-old Francis Ormond Jonathan "Fog" Smith of Maine, was in a receptive mood. Morse proposed that his telegraph be placed under government control to thwart speculators and monopolists. Fog Smith, too, wanted the telegraph to fall into the right hands. In Smith's view, they were the hands attached to Fog Smith.

Not much is remembered about Fog Smith, although he alternately entertained and infuriated his contemporaries for years. Like Morse, he was a New Englander, born in New Hampshire, and in the small world of the early nineteenth century he was, as it turned out, Morse's distant cousin by marriage. He had followed his tavernkeeper father north to Maine, arriving, according to Morse's definitive biographer, Carleton Mabee, without five dollars or five grains of integrity in his pocket. But he had a winning way with him. He was admitted to the bar, got elected to the Maine legislature at twenty-four, became president of the state senate in two years, stood for Congress, and won. Smith, wrote Benjamin French, the clerk of the House, "is without exception, the most driving, persevering, energetic man I ever knew. You will always find Mr. Smith a warm-hearted friend. When he *takes*, he makes no half way business of it, & *vice versa* — he is, what Dr. Johnson would call 'a good hater.'"[5] At the time he met Morse, he was under investigation in Massachusetts for his banking practices. But Morse saw nothing out of line in Fog Smith, and neither did Vail or Gale. He was clearly a wealthy man — he was building a mansion in Maine — and the Morse partners needed money. He was a lawyer, and they didn't have one. And he was a powerful congressman and the chairman of the very committee that would have to pass on the telegraph. So they cut him in.

The group that would later be known as the Morse Patentees was now complete. Morse retained a nine-part share of the telegraph in the United States and eight parts abroad, Smith was given four parts and five parts, the Vails held two parts, and Leonard Gale, who was the only one of them who knew anything about electrical theory, had one part. Vail and Gale, designated as the technicians, would do the work, Morse and Vail would pay the partnership's American expenses, Smith would represent them as a lawyer and foot the bill for a three-month trip to Europe that he and Morse planned by way of testing the waters for the invention.

In America, things were looking good. Official Washington seemed wonderstruck by the electromagnetic telegraph. In an age of proliferating inventions, all things seemed possible, the telegraph was the latest of them, and the government had yet to face the prospect of putting up some money for it in the middle of a financial crisis, something that invariably focuses the governmental mind. Senator John C. Calhoun was sufficiently impressed to send a delegation of his colleagues from the upper house to view the amazing machine. With his cabinet, President Van Buren had also viewed it and dictated a message prophetic of his doomed hopes of future tenure: "The enemy is near." Fog Smith, tactfully neglecting to mention that he was now an owner of the telegraph, sent a glowing report to his committee. The time had come, as the partners had planned, to sell the telegraph in Europe. Morse, who had filed a caveat announcing his invention with his old Yale classmate, the chief of the Patent Office, postponed his actual application for a patent in the belief it would interfere with the European sales effort. It was a small mistake with no real consequences, but it contributed a minor, little-known footnote to history. As a result, the first electromagnetic telegraph patented in America would be British. It was the telegraph of Wheatstone and Cooke.

* * *

The Europe Morse had visited as an artist between 1829 and 1831 had been full of line-of-sight telegraphs. The Europe the telegraph entrepreneurs Morse and Fog Smith visited in 1838 was full of electromagnetic telegraphs.

Cooke and Wheatstone were building a line in Belgium, and they planned to string two hundred miles of wire in England. Edward Davy, a London apothecary and no relation to Humphry Davy, also had built some sort of telegraph, and he was confident enough in his invention that he had challenged the Wheatstone patent and received (through the intervention of Faraday) a patent of his own. This was not good news. Davy, Wheatstone, and Cooke all contested Morse's own patent application, and the British attorney general, Sir John Campbell, through whom the path to the patent lay, was not inclined to look with favor on the American device. Morse, cooling his heels and getting nowhere, stopped by the exhibition rooms in Exeter House where Davy's telegraph was on display at the cost of a shilling. He was not impressed. It was a needle device with a keyboard, as complicated as the Wheatstone system he had seen described in the scientific papers. Morse concluded that his own device had the considerable virtue of simplicity. But Davy had also invented a relay, an electromagnetic switch that allowed many telegraph receivers to be hooked into a line rather than only two of them. Henry would later become extremely interested in the question of whether Morse had seen a relay in Europe.

Then, hanging fruitlessly around the attorney general's office, Morse met Wheatstone, who invited him to inspect his version at King's College. Morse was quite taken with the young scientist, and he wrote his brother that Wheatstone's telegraph "is truly an ingenious and beautiful piece of mechanism, but" — he added — "it is not so simple as mine."[6] In a fair fight, he was certain his telegraph would prevail. And it did. Davy's device went nowhere after he fled to Australia to escape his creditors, and the Wheatstone-Cooke telegraph was largely confined to the British Isles. But Morse's inspection of the British systems is interesting for another reason.

There has long been the question of what, if anything, Morse actually invented. He was using Henry's electromagnets and he would later steal — legally, in those days before cross-border intellectual property agreements — another and better one from the Frenchman Louis Breguet. Guided by Leonard Gale, he was using the new English batteries of Cruickshank and Daniell. Alfred Vail had devised the key. Indeed, the only element of the Morse telegraph to which Morse could lay claim was the relay that alerted the stations along a telegraph line to the fact that a message was coming through.

As Gale remembered it, he had pointed out to Morse that as a practical matter the electric telegraph couldn't operate at distances longer than twenty miles. The problem was, once again, Ohm's Law, this time combined with the laws of induction. A weak electrical signal, intensified by the proper array of batteries, would travel inexorably down any conceivable length of line until it completed its circuit. Thanks to the laws of induction there was no stopping it from going where it wanted to go as long as the system was self-contained, because the circuit consisted of two stations at the ends of a single line. But a commercial telegraph system, by definition, could not consist of a single line with only two stations. There would be branch offices, the branch offices would receive messages, the resistance in the receiving apparatus would further weaken the intensity of the current, and eventually the falloff would prevent the current from doing work.

According to Gale, Morse replied to this rather considerable objection by saying that an electromagnet could be used to break the circuit — that, in fact, a series of magnets could be used to break the circuit at a multitude of points "and so on around the globe." Perhaps Morse actually did say something like that, although Gale, like the Vails, had a vested interest in keeping Morse's inevitable patents pure of any hint that something pertaining to the telegraph had been invented elsewhere by somebody who was not Morse, as Henry suspected. Morse's alleged remark seems to describe, although vaguely, something called a make-or-break switch, a switch that would interrupt the main circuit and direct the signal to the branch office — but here the partnership agreement rears its head. All inventions and improvements to the system, no matter who actually made them, had to be assigned to Morse. And Morse knew very little about electricity — so little, in fact, that his vague remark, if he made it, appears to be nothing but a commonsense hint at a usable solution that he had no idea how to build. Something a little more complicated than a magnet was required.

A relay was needed. It would work like this: At a branch office, a make-or-break switch powered by an electromagnet would be opened by the weak current from the main line, and the weak current would then activate a local receiver powered by its own local battery. This would relieve the weak current on the main line of the burden of sending the message to the branch office. The local receiver at the branch office would then clatter out the message using its own power source on the branch line, and the weak current on the main line would follow the laws of induction to the main line's end, while the copy of the message that reached the branch terminated there. And this could happen as many times as there were branch offices.

This was, in fact, the solution that was adopted. It was not mentioned in Morse's caveat — a warning that he was working on an invention, but not an actual application — to the Patent Office in 1837. A reference to a relay appears in the revised specifications he filed in April 1838, just after he met Joseph Henry in person and just before he departed for England and Europe. Even then, according to Morse's principal biographer Carleton Mabee in his *American Leonardo*, "there is evidence . . . to show that SFBM did not know its importance and was not sure of the necessity of using it even then." Henry later testified that he didn't learn of Morse's relay system until 1839, a year after Morse had returned from Europe. In London, Morse had inspected both the Wheatstone and Davy telegraphs. And both of them incorporated make-or-break relays. "I am morally shure," Henry later wrote in the margin of a book commemorating Morse, "that Morse had no k[n]owledge of a local magnet until he went to London in 1838."[7]

Otherwise, the trip to Europe was without fruit — so fruitless, indeed, that Fog Smith gave up and went home. Sir John Campbell continued to deny a British patent on a technicality, and the only recourse was an act of Parliament, a time-consuming process the Morse partners could not afford.

In France, the country still loved its line-of-sight semaphores, but Morse was hailed by François Arago, the director of the Paris

Observatory, and spoke before the Academy of Sciences. There was already interest in the Steinheil telegraph from the Germanies, the French knew of Wheatstone's 1837 telegraph, and Morse had no trouble gaining a patent. It came with a catch: if the patent was to be valid for more than two years, he had to build an actual telegraph. Morse set up his implements in the rooms he shared with a Philadelphia expatriate named Mellen Chamberlain, demonstrated his telegraph to many enthusiastic visitors, and was delighted when Chamberlain signed a contract to peddle it in Prussia, Austria, and the Near East. Bavaria was out: Steinheil had already agreed to build a telegraph system there. But before anything could happen in the East, Chamberlain drowned in the Danube, taking whatever agreements he had closed with him.

For a time, the Rothschild interests seemed inclined to include a Morse telegraph in the seven-mile railroad line they were building from Paris to Saint-Germain, but nothing came of it. With time on his hands and the clock ticking on his patent, Morse made the acquaintance of Louis Daguerre, whose new invention of photography was the rage of Paris. While they met in Morse's rooms, Daguerre's studio burned down. Still, Morse was deeply intrigued by Daguerre's invention. When he returned to New York, he became one of a handful of people who introduced photography to America, opening an experimental studio on the roof of one of the NYU buildings and also opening a school. One of his pupils was Matthew Brady, the great Civil War photographer.

Morse and his telegraph, too, were a Parisian rage, but he was getting absolutely nowhere. Then the Russians came forward in the person of Baron Meyendorf, and it seemed that the day was saved. Meyendorf was prowling Europe in search of useful inventions. He had already struck a deal with a French inventor, a certain Monsieur Amyot, to build a telegraph line for his tsar, and if Morse and the Frenchman could agree to terms, Russia would become the first country in the world to adopt the Morse telegraph system. Morse and the Frenchman found that their interests coincided, Meyendorf held out the prospect of travel expenses so that Morse could supervise the construction of the twenty-mile line in person, and he returned to St. Petersburg to report to his royal master. Morse returned to America full of plans.

The tsar, unfortunately, decided that he didn't need a telegraph. At the end of it all, there was no hope in England. Morse had been lionized in France, but there was no hope there, either. That left America, back where he had started.

* * *

But in America, Congress was no longer much interested in telegraphs. Andrew Jackson was long gone from the White House, but his mad economic depression raged on and on, putting a serious crimp in the legislators' eagerness for new adventures on the public nickel.

As for Morse, he was no longer very interested in painting. His brushes dried out and his canvases gathered dust. His purse was even shorter than usual, and he railed at his students about the folly and penury of the artist's life. Before his trip to Europe, he had begun research for his *Mayflower Compact*, but although it was bought and paid for, he was no longer interested in that either; he promised to repay the money, although he had none. Help was no longer forthcoming from the Vails as the judge had unwisely invested in the new railroads at the wrong time, and Alfred was working at the company that would soon become the Baldwin Locomotive Works. Fog Smith, landing on his feet as usual, had returned to Maine to edit a farming magazine, but he still regarded himself as cut in on any deal Morse could close. Leonard Gale resigned from NYU during a faculty protest and took a teaching job in Mississippi, but not before he loaned Henry five miles of Morse's telegraph wire so Henry could conduct some more electrical experiments. This — and the timely arrival of an installment of Henry's scientific papers — gave Morse his opening.

He appealed to Henry for help.

To the modern eye, Morse's letter was obsequious, overwritten, and fawning; to his contemporaries, it was a letter from an old Phillips Andover boy, Yale scholar, and gentleman seeking a favor.

And Henry, with the spelling and punctuation of his reply as innovative as always, was the accomplished and cordial professional putting an amateur at ease. He thanked Morse for the loan of the wire, and he invited him to Princeton. And, in contrast to the discouraging advice he had given Thomas Davenport and choosing words Morse could show to Congress, he wrote, "I am acquainted with no fact which would lead me to suppose that the project of the electro magnetic telegraph is impractical, on the contrary I believe that science is now ripe for the application and that there are no obvious difficulties in the way. But," Henry added in words that would return to haunt the two of them, "what form of the apparatus or what application of the power will prove best can only be determined by careful experiment. I can say however that so far as I am acquainted with the minutia of your plan I see no practical difficulty."[8]

Henry was perfectly aware that there were other electrical telegraphs in the world, and he either knew or suspected where Morse had gotten the components for his system. He knew there was nothing new in there. Moreover, he would repeatedly say so even as Morse tried to insist that the others had stolen his designs and that he had no idea that Henry had described the components and physics of a working telegraph in 1831. The bitterness between them, however, lay in the future, when Morse went to court to defend his patent in the 1840s and flatly denied that Henry had made a contribution; at the moment, Morse had no idea that Henry, good scientist that he was, expected him to acknowledge his debt to others, including Henry himself.

Congress was unimpressed with Henry's endorsement, and Morse, as always, found distractions elsewhere. He made his second disastrous run for mayor. He took the second photograph ever made in America. But he kept his eye on the main chance, and the main chance was still the telegraph. On June 20, 1840, he finally obtained a patent. Unfortunately, Wheatstone and Cooke had obtained their own American patent only eight days earlier, and to them went the pride of place in owning the first officially recognized telegraph in the country.

Morse strung an underwater wire insulated with pitch, tar, and rubber from Governor's Island in New York harbor to Castle Clinton on the Battery, almost the very tip of Manhattan. The *Herald* announced that a revolutionary new invention would be demonstrated to the public, but a ship's anchor cut the cable before the exhibition was made. Next, unaware that a British army surgeon had already done it in the 1830s by using the Hooghly River to complete the circuit on his own telegraph in India, Morse used water itself as a conductor and sent a signal across a canal. Congress remained unimpressed, and Morse remained uninterested in any customer but Congress.

Henry continued to lend his support. "I am pleased to learn," he wrote on February 24, 1842, "that you have again petitioned congress in reference to your telegraph and I most sincerely hope that you will succeed in convincing our Representatives of the importance of the invention. The science is now fully ripe for such an application, and I have not the least dout, if proper means be afforded, of the perfect success of the invention."[9] Of course, he added, Professor Wheatstone and Dr. Steinheil also had their telegraphs, but "unless some essential improvements have been made in the European plans I should prefer the one invented by yourself." Morse took the endorsement to heart, but he failed to note a certain implication in Henry's words. The scientist had hailed Morse as the inventor of a superior kind of telegraph. But Henry did not recognize him as the inventor of the telegraph itself.

Morse was back in Washington again, carrying on the fight alone with Henry's distant and still cordial support. It was the custom of the time to give a promoter of a project a room in the Capitol where he could display his wares, and Morse received two of them, one at the House Committee on Commerce and the other at the Senate Committee on Naval Affairs. He strung his wires between them, set up his transmitter and receiver, and invited in the legislators for a demonstration. It was not the most inspired venue.

For the first time, the congressmen could examine the telegraph closely and at their leisure. At Morse's suggestion, they gave him messages to transmit. He worked Vail's key, after a moment his assistant appeared in the doorway with the text, and Morse triumphantly read it off. But a large number of congressmen — a dangerously large number — had no idea what Morse had done or why it was important, and some of them concluded that he wasn't quite right in the head. "I watched [Morse's] face closely to see if he was not deranged," reported Senator Oliver Smith of Indiana after an early test, "and was assured by other Senators as we left the room that they had no confidence in it either."

In the early 1840s, Senator Smith and many of his colleagues simply had no way of understanding what the telegraph was or how it worked. Because everybody had boiled water, a steam engine was easy to understand, and the work it did was the sort of work that had always been done, only more powerfully and much faster. Because people had been melting rocks for thousands of years, an iron mill was not a mysterious place. The Industrial Revolution was easy to understand, because it allowed people to do what they had always done in ways that were more efficient and easily comprehensible. But there had never been anything like the telegraph in all of human history. It did something that people had always done — convey messages from place to place — but in a way that was radically better, because the transmission was instantaneous. Never in human history had there been anything like it; for eons, messages had traveled at the pace of a sailing ship, a trotting horse, or a walking messenger. To comprehend the telegraph required a revolution in human thought, and not everyone was immediately capable of it. According to the laws of physics as they understood them, it was impossible. Certain congressmen thought Morse was mad because of what he was doing.

In the final telegraph debate in the House on March 3, 1843, Congressman Cave Johnson suggested, amid laughter, that the Morse subsidy, if passed, should be shared with the mesmerists — the followers of Anton Mesmer and his theories of animal magnetism, which a panel of scientists that included Franklin and Lavoisier had debunked in the Paris of the 1780s. Others proposed adding the Millerites, the followers of an apocalyptic religious sect founded by Williams Miller in 1831, who believed that the end of the world was nigh. When the favorable vote finally came, granting Morse $30,000 for a line between Washington and Baltimore, it was by voice rather than record, because many of the legislators were reluctant to attach their names to a preposterous scheme that would waste such a titanic amount of the taxpayers' money. Seventy congressmen did not vote at all rather than brave the voters' wrath. Representative David Wallace of Indiana voted for the subsidy and was not returned to office for that very reason.

Passage in the Senate went down to the wire — if it failed, Morse was worried that he wouldn't have enough money to buy a return ticket to New York after he settled his hotel bill; he had thirty-seven and a half cents to his name — but the vote there, too, was favorable. Now all Morse had to do was string a telegraph wire from Washington to Baltimore.

He had no idea how to do it.

* * *

With the subsidy in hand, he was able to pay himself a salary as the official superintendent of United States telegraphs. He was also able to reclaim the services of Vail and Gale and actually give them wages. Fog Smith, rusticating in Maine, was not included in this happy arrangement, but once Morse had decided to lay the line underground — because Wheatstone, in England, had originally laid his line underground — and put the contract out for bids, behind his back Smith rigged the auction, appointed a front man, and became the contractor himself. And it was thanks to Smith that Ezra Cornell became the last valuable addition to the Morse team.

Ezra Cornell was a mechanic from Ithaca, New York, who had a convenient sideline selling plows, a job that had brought him into contact with editor Smith of the *Eastern Farmer*, who gave Cornell's sideline a pleasing amount of publicity. It was, in fact, a plow — of sorts — that Smith was kneeling beside on the floor of his office when Cornell walked in during the summer of 1843. It was Smith's very own invention, and it had something to do with something called a telegraph line. It was supposed to dig a trench and leave the earth heaped up on either side. A second machine would then come along and fill the earth in. Unfortunately, it didn't seem to work very well. Cornell designed a better one, a machine that would both cut and cover, and soon he and his new plow joined Morse in Washington.

Everything was ready. Morse, using the new Grove batteries, had successfully sent a current through 160 miles of wire. The signal would actually reach Baltimore from Washington. Permission had been obtained from the Baltimore & Ohio railway to run the wires alongside the tracks, the lead pipe for the underground conduit was stacked and ready with the insulated wire enclosed, Cornell whipped up his team, and the job was begun — just forty days before the agreement with the government said it was supposed to be finished. From the beginning, nothing went right.

The insulation in the pipes failed; Cornell had suspected it would. The wire had been wrapped in cotton twine and sealed in a vacuum in a lead pipe with a half-inch diameter. Perhaps the vacuum had failed or been imperfect. Whatever the reason, the wire had failed, miles of underground telegraph line were useless, and the clock was ticking. Smith, who had appointed himself general contractor, demanded to be paid in full for a job that was not completed in full and in fact was not even properly begun. He threatened to sue when Morse refused to see his version of reason, and righteously told the secretary of the treasury that Morse was a very corrupt fellow. Only $7,000 of the government appropriation remained, not one inch of the line was usable, and the government contract was running out. To gain time, Cornell sabotaged his pipe-laying machine. Crying, "Hurrah, boys, we must lay another length of pipe before we quit," he ran it into a rock. Without the pipe layer, no work could be done. "*Hinc illae lachrymae,*" Morse wrote. All is sorrow. A good Yale man, he was quoting Horace.

Although Morse had written vaguely in his patent application about "pillars" that might support the telegraph line, it was Vail who came up with the idea of poles. He and Cornell were reading every scientific paper they could lay hands on in the Congressional and Patent Office libraries in Washington, and discovered that Morse had copied Wheatstone all too well. The insulation on the English scientist's buried line had also failed. Now it was again time to copy Wheatstone, who had next strung his wire on poles. At Vail's urging, Morse put the contract for the poles out for bid on February 7. The new poles were debarked chestnut, placed twenty feet apart. Not to be upstaged, Cornell salvaged the wire from the buried and now useless line, reinsulated it with more cotton twine soaked in shellac, and used glass doorknobs to insulate the wire from the wood, which would sometimes be wet and, in the absence of the additional insulation, short out the line. Henry had suggested glass long ago.

Less clear is the identity of the person who came up with the idea for the earth return. The earth return grounded Morse's system and cut in half the amount of wire he would have to use. For electricity to move through a wire, it must move in a circle: in order to reach its destination — Baltimore, for example — it must then flow back to its point of origin and begin the journey again. Building an electrical system that does not allow the electricity to flow in a circle is like building a ditch with no outlet and no inlet. In such a ditch, the water will simply stand. So too with electricity. If electricity does not move in a circle, it too will simply stand and not move, and no message can be sent. The early telegraphers like Morse used two wires for each transmission channel, one to carry the message to its destination and other to return the electricity home. However, experimenting with the use of railroad tracks as an electrical conduit, Karl August von Steinheil, the Munich astronomer, had discovered the principle of the grounded line. If a telegraph line were grounded, the earth itself would return the current — not very efficiently, but well enough — and the second wire could be eliminated. Morse had apparently heard something of Steinheil's discovery when he was last in Europe, but as always with Morse, there is the question of when he had heard it and the second question of what he had understood. It was Vail, not Morse, who was doing the necessary library work during the crisis, aided by Cornell. And it was Vail who submerged a copper grounding plate near the Pratt Street docks in Baltimore and buried another one in the Capitol in Washington. The cost of Morse's wire fell by half.

On May 24, 1844, Morse kept a promise to Annie Ellsworth, the young daughter of his old Yale classmate who had become the patent commissioner. She was to pick the first momentous message to travel over an (American) intercity telegraph line, and she chose her text from the Book of Numbers. Transmitted by Morse from the old Supreme Court chamber in the Capitol to Vail in Baltimore, it was: "What hath God wrought!" But the legislators were still unimpressed.

They were unimpressed for a couple of reasons. They had been keeping close watch on Morse's operations and the fate of the staggering $30,000 they had given him, and Morse, Vail, and Cornell had been telegraphing up and down the line for weeks. So jaded had the legislators become, in fact, that on May 1, when Vail had telegraphed from Annapolis Junction that the Whig national convention in Baltimore had nominated Henry Clay for president and Theodore Frelinghuysen for vice president, the general reaction was not "My God, it works!" but "Who the hell is Frelinghuysen?" It was easy to check the accuracy of the message. An hour later, the same message arrived by rail. Yes, it was Frelinghuysen, whoever the hell he was.

When it came to the telegraph, Morse was one of the luckiest people who ever lived. The Democrats were also holding their national political convention in Baltimore, and Morse and Vail were placed to provide, in real time, the politicians of Washington with their favorite fare, politics, as the dark horse Democratic candidate James K. Polk came roaring in from nowhere and seized the nomination. As Vail's bulletins from his improvised telegraph office reached the Senate floor, pandemonium finally reigned. "The gavel thundered," Ezra Cornell recalled, "and the lightning continued its flashes until the storm burst with such fury that an adjournment was moved and carried." No one ever again doubted that the telegraph would have its uses. Every nation has its obsessions. The British fascination with crime made Wheatstone's telegraph a success after a long period of public and corporate indifference. Similarly, the American obsession with democratic politics was the making of Morse's system. The same could not be said in France, still wedded to its semaphores, or the Germanies, which were actually more advanced in the telegraph line. In France and Germany, unlike England and America, there was little popular enthusiasm for the telegraph because no nationally fascinating telegraphic event had occurred.

Thanks to politics, Morse's name and his fortune were made. The legislators had been uninterested in "What hath God wrought!" but the unexpected nomination of James K. Polk had definitely captured their attention — and the attention of the press. But if Morse's name and fortune were made, the same could hardly said of his extremely competent helpers, with the possible exception of Fog Smith and Ezra Cornell.

* * *

What had Morse actually done? The electromagnets were Henry's, the poles were Wheatstone's, the concept of earth return was Steinheil's, the origin of the relays was suspicious, the improved batteries were Grove's, the equipment was built by Vail, and Morse's leadership abilities left a great deal to be desired. "Professor Morse is so unstable and full of notions," Vail wrote after the wire crisis had been resolved. "He changes oftener than the wind, and seems to be exceedingly childish sometimes. Now he is elated up to the skies, and then he is down in the mud all over under. It requires the utmost patience to get along with him." Had Morse, then, made nothing but a lucky guess and somehow carried through despite himself? Not quite.

In a single stroke, long years ago aboard the *Sully* in 1832, he had conceived a telegraph that was elegant in its simplicity, easy to use — it did not involve the reading of Wheatstone's complicated message board — and able to record the messages it received. Later, to his regret, telegraph operators would learn to read the code by ear as the relay opened and closed and the electromagnet-powered receiver clattered away — Vail revised a new receiver that capitalized on this — but the change did nothing to alter the essential beauty of the original design. Although he studied the European needle systems that, operationally, were far in advance of his own, he never changed his system. Morse was in many ways a weak, morally dubious, and distractible man, but he persevered for years against the indifference in Washington. The telegraph of Wheatstone and Cooke was never widely adopted outside England, other European needle telegraphs fell by the wayside, and the Morse device and the code that bore his name became the global standard for a hundred and fifty years, until the last marine telegrapher signed off in the 1990s and the system went dark forever, finally superseded by satellite communication. He did not invent the telegraph, but he had conceived of a better one and he foresaw the Atlantic cable. None of his associates were capable of any of this.

Leonard Gale was an unassuming and conspicuously unambitious scientist. After he and Morse parted company in the mid-1840s when Gale left to take up his new teaching position in the South, he surrendered his share of the partnership. His contributions were many, but no one remembers him now. Ezra Cornell was a highly competent mechanic. He became wealthy from his later investments in telegraph lines and founded the university that bears his name in his hometown of Ithaca, New York. Fog Smith, the politician and low-level scoundrel, built a telegraph line of his own from Boston to Halifax, Nova Scotia, bought a house in the fashionable Williamsburg section of Brooklyn for his new wife, and left a will, remarkable for the period, in which he named his illegitimate children. He, too, is nearly forgotten. Of all the Morse associates, only Alfred Vail and Joseph Henry have reputations that live on in the shadow that Morse never really deserved to cast.

In 1845 Vail published a book, *The American Electro Magnetic Telegraph,* with Morse's blessing. It appeared to be the definitive text on the new invention, and Henry's name did not appear. Henry was not pleased. There were limits to his generosity, and Vail had challenged them.

"HENRY sticks it into MORSE," John Buhler, a young Louisianan attending Princeton, reported to his diary in early 1846. "Says M's assistant *Vail* has lately published a book — purporting to be a history of *The Telegraph* & hasn't mentioned him atal in it — although it was through communications & instructions freely made by him — that M's telegraphic scheme came to a consummation."[10]

Henry, unfortunately for his relationship with Morse, possessed an ego. As a dedicated scientist, he was always careful to give credit to others where credit was due, but Vail's enormity in leaving him out of the first book written about Morse's telegraph went far beyond a question of attribution. Without Henry's electromagnets, Morse's telegraph would not work, and without the insights in Henry's 1831 paper, Morse couldn't have built a telegraph at all. Like Cooke in England, Morse and his vision succeeded only when he came into contact with someone who understood electricity. For years Henry had demonstrated the principles underlying the Morse telegraph before dozens of his pupils. Now Morse and his cat's paw were assigning the credit solely to himself, a scientific ignoramus. It mattered not a jot or a tittle that Vail had signed a document requiring him to do just that, and that Vail himself never mentioned his own considerable contributions to the project. Henry may have known nothing of the partnership agreement, but it wouldn't have mattered if he had. Scientific truth was being distorted beyond recognition. In a word, Vail — and probably Morse — was lying.

Morse soothingly insisted that the book was Vail's, not his, although Vail could not have written it without his blessing. He said that he hadn't read the offending pages — which, given Morse's ego, was unlikely. He promised to have the oversight corrected in the second edition. When a second printing of the book was struck from the same plates and the omission persisted, Morse said it wasn't a second edition. Then his lawyers attempted to destroy Henry's scientific reputation.

In the process of invention, there are typically three phases. When the inventor first proposes to make his invention, nobody can do it. When he makes his invention, everybody could have done it. And when his invention begins to make pots of money, everybody did it before he did. Claude Chappe, the inventor of the French optical telegraph, had been unhinged by this inevitable and remorseless process, and he killed himself by jumping into a well outside his telegraph headquarters. Morse himself had also had a taste of it. Dr. Charles Jackson, the young physician whose remark had started Morse down the telegraphic trail on the *Sully*, announced that he was the coinventor of the electromagnetic telegraph. Then Jackson decided that he had invented it by himself, and he made a second public announcement to that effect.

It was true that Morse had continued to have a distant and casual relationship with Jackson. They had exchanged one or two letters. It was also true that they had spoken together of a telegraph on the ship, with Morse, as Jackson pointed out, initiating the conversations, and that Jackson had independently dabbled a bit in that direction, although without much in the way of results. But Jackson, until Morse made his first proposal to Congress, did not regard an electric telegraph as a particularly useful device. And it was true that Jackson was an undeniably brilliant man. For a brief span of time, a few people believed that he probably did have a hand in inventing the telegraph. He had stopped practicing medicine, and he had set himself up in Boston as a chemist and geologist. At the time he first made his telegraphic claim, he was, in fact, the state geologist of Maine, an office he would also shortly assume in New Hampshire. A mountain in New England still bears his name. He became a professor at the Medical College of Massachusetts, the school that later became Harvard Med, and after the Morse incident, he developed an odd habit of claiming other people's inventions as his own. In 1846 he put himself forward as the man who discovered the surgical uses of anesthesia in the form of ether. It was true that he had experimented with it, but the first person to use it in actual surgery was a dentist named William Morton. Jackson also claimed that he, and not C. F. Schönbein, had invented guncotton, a powerful explosive made from nitrocellulose. But Jackson, who eventually died in a mental institution, presented Morse with a minor, if bizarre, kiss of the lash that was to come. Once it became clear that Morse's telegraph was likely to bring him wealth, to say nothing of fame, there seemed to be rival telegraphs everywhere, with Henry, the nation's premier expert on electricity, speaking on their behalf.

Not only did Morse claim to be the inventor of the electromagnetic telegraph, but he had a patent to prove it — a patent that he briefly extended to cover all electromagnetic telegraphs of whatever type that were introduced anywhere in America, although it soon collapsed of its own absurdity. He had tried, in effect, to patent electromagnetism itself. He successfully challenged the patent of Royal E. House, a Vermonter who had also invented a gate that could be opened without alighting from a vehicle or horseback, and who had devised a short-range telegraph that printed its messages in Roman letters. He challenged the telegraph of Alexander Bain, a British clockmaker, whose high-speed telegraph printed its messages, using a component that resembled Morse's portrule, with an electrochemical reaction on paper soaked with potassium iodide — the forerunner of the fax machine. Testifying for the defense, Henry calmly pointed out something he had repeatedly said, that while Morse's telegraph was a clever solution, there was nothing whatever that was new or original about anything in it.

Here was grave danger. Morse clearly believed that if Henry was believed, his patent would be overthrown — and on the face of it, although it was less true in practice, the Morse patent was a license to coin money. The money machine, when it was in full working order, chugged along like this: Once the telegraph was privatized, the cost of a single-wire line was usually around $150 per mile. Perhaps $75 more went to the line's promoters, the people who boomed its stock and sweet-talked the owners of the right-of-way. The cost reported to the stockholders was $600 per mile, which left $375 in the kitty of the line's management. Morse and his fellow patentees — Smith, Gale, and the Vails — claimed half of that sum and sometimes got it; management had various dodges and excuses, which occasionally worked. But in theory, there was a mountain of gold out there, and Henry was a clear and present danger to the mining of it. Morse's lawyers lost no opportunity to attack him.

In the end, Morse and his enemies were wrong about the patent. Rendering the majority decision for the Supreme Court in an 1850 case that challenged Morse on the grounds of unoriginality, Chief Justice Taney carefully reviewed the history of the telegraph, cited the contributions of Wheatstone, Davy, and Steinheil, generously acknowledged Henry's contributions to the science of electromagnetism, failed to mention Alfred Vail in any context, and determined that Morse had a perfect right to use the discoveries of others because "otherwise, no patent, in which a combination of elements is used, could ever be obtained." The decision became the basis of American patent law. It also had a profound effect on the thinking of Samuel F. B. Morse.

In 1832 Morse had believed he was the only person on earth who had ever conceived of an electromagnetic telegraph. Now he believed it again. Earlier, he had vigorously tried to mollify Henry by recognizing his achievements, but he was no longer in a mood to share even a fraction of the credit. In the late 1840s and early 1850s, in a deliberate attempt to ruin Henry's career, he called him a liar. "I am not indebted to him," he wrote, "for any discovery in science bearing on the telegraph."

Henry, the calm man of science who was by then the first secretary of the new Smithsonian Institution, easily withstood Morse's attack. He became Abraham Lincoln's friend, and he founded the predecessor of the National Weather Service, an organization that was made possible by the telegraph. Port Henry on Lake Champlain was named for him. When he died in 1878, thirty years after Morse's attacks, the president and the cabinet attended his funeral. He never forgave Samuel Morse.

Alfred Vail did not fare so well. Restrained by the contract his family had signed, he never claimed credit for the huge contributions he made, the code, the key, and the redesign of the rest of the original equipment, and he was never able to patent them. He continued to run Morse's telegraph for a while, and he spent the rest of his short life compiling a Vail family genealogy. Fort Monmouth, New Jersey, the headquarters of the Army's Signal Corps, was originally named Camp Alfred Vail.

Late one night in 1911, a descendent — believed to be a grandson — crept into the graveyard where Vail lay buried and added to his tombstone the line INVENTOR OF THE MORSE CODE.[11] A cousin, Theodore Vail, led AT&T into the long-distance telephone business that eventually helped destroy the telegraph as a means of communication.

* * *

Morse himself became financially comfortable for the first time since leaving Charleston in the 1820s, but he never became wealthy by the standards of the robber barons into whose age he lived. When it was discovered that J. P. Morgan had left an estate of $70 million, E. H. Harriman burst out: "And to think! He wasn't even a rich man." Morse could have entered that league — for a time, he controlled the telegraph in most of the United States — but he never did.

At first he tried to sell the telegraph to the government for $100,000, but the government was preoccupied with the war in Mexico and no further money materialized from that quarter. Morse then engaged the services of Amos Kendall, Andrew Jackson's old postmaster and biographer, who established the Magnetic Telegraph Company as the vehicle for directly commercializing Morse's invention. Indirectly, of course, Morse and his partners tried to collect their fifty percent as other companies entered the business, usually with no more knowledge of the technology than Morse had. In the early days, telegraph service in America was atrocious, especially after a storm. Work crews had no idea of proper pole placement and sometimes stapled the wires to living trees. They did not understand insulation; the sins of omission and commission were many. Morse, when he was called upon to testify in court in defense of his patent, in the 1850 case and on other occasions, revealed that he still didn't understand much about electricity, either.

Morse never painted again. In 1847 he built a house near Poughkeepsie, named it Locust Grove, married an impecunious, speech- and hearing-impaired cousin thirty years his junior, and raised a second family to go with the children he had from his first. None of his progeny distinguished themselves. He served briefly as a consulting electrician to Cyrus Field when Field was laying the Atlantic cable, but as a technology, the telegraph did not sustain his interest. He was content to collect his royalties until his patent expired, and he was happy to let his countrymen hail him as the man who had taught the lightning how to talk, trotting himself out for ceremonial occasions where he conspicuously failed to remember the names of any of his telegraphic colleagues and advisors. When the nations of Europe gave him an award of 400,000 francs, he neglected to mention that they had voted a like sum for Steinheil. As he aged, his fellow countrymen bestowed upon him the elder statesman role they had chosen for their electrical pioneers. Like Franklin before him and Edison after, he came to be viewed as a benevolent old man. It was a role he found congenial. On April 2, 1872, he died at the age of eighty-one from complications of a cold contracted during yet another ceremonial address in which he mentioned no names but his own.

* * *

At first it seemed as though almost no one actually wanted to *use* the electromagnetic telegraph, however much they were fascinated by it. It seemed as though it would never catch on, much less catch fire. On the inaugural Baltimore-Washington line, the price of a message was four alphabet characters for a cent. A cent, as it happened — just one — was the entire revenue for the first four days of operation. Receipts for the first week were a dollar. People would crowd the office to see the apparatus — and to say they had seen it. Other companies were soon formed to exploit Morse's invention — and there were many, including the one founded by Hiram Sibley of Rochester, New York, that became the predecessor of the Western Union Telegraph Company that he formed in 1856. Like their congressmen, the Americans were uncertain what to make of this new thing, and like the capitalists stringing the new lines in all the wrong ways, they had only the most rudimentary and often mistaken idea of how it worked.

The Americans of the 1840s were experiencing a form of future shock. Thanks to the railroads just then being built, they could physically travel at unprecedented speed over distances that had once seemed immense, and so could the produce of the farms and the products of the factories that were increasingly powered by steam. Once a certain amount of mental inertia was overcome, so could their messages, the business correspondence of their employers, and the train-routing instructions of their railroads. No longer did a businessman in New York meet with his suppliers twice a year, usually in the spring and the fall. Now he could communicate directly all year long and fine-tune his inventory. Indeed, members of the business classes began to complain that they now worked all the time and had no leisure. Speculators no longer had to physically visit the stock exchange to follow and exploit the market. They could station agents there to signal them of opportunities via the telegraph, and later they could own their own telegraph-based stock ticker that followed the entire market all the time. Whole cities no longer burned down because a fire raged unknown until too late; telegraphic fire alarms alerted the engines from hundreds and thousands of street corners. The telegraph was everywhere — sometimes, as in the case of the stock ticker and the fire alarms, invisibly — and people made it part of their lives.

By 1851 there were twenty thousand miles of telegraph lines in the country, owned by fifty different companies whose systems did not connect. These were the companies that were combined by Hiram Sibley to form Western Union, bringing order out of chaos at the price of near-monopoly control, and in 1861 the first transcontinental telegraph was strung to San Francisco, driving the Pony Express out of business after only a year and ruining its investors. In 1867 the United States bought Alaska from Russia, in part to obtain ownership of a vital link in around-the-world telegraphy, a development that it was believed would bring world peace as the war-prone nations of the earth began to talk to each other in real time. It was a delusion that would attach itself to every major advance in electrical science thereafter. Instead, something else began to occur. In 1869, when the tracks of the first transcontinental railroad were joined at Promontory Point, Utah, the blow of the hammer on the golden spike caused a telegraph signal to be sent that simultaneously fired cannon on both coasts to mark the occasion. With Morse's invention, the first ubiquitous electrical device that performed useful work, the applications of electricity began to mutate in ways that were logical but could never have been foreseen.

* * *

It has been repeatedly written that the telegraph forever liberated human thought from the constraints of time and distance, but this is more than a little disingenuous. Before the electromagnetic telegraph arrived on the scene, many hundreds of thousands of people were communicating quite nicely and with remarkable speed. True, the business classes found that they no longer enjoyed the twin luxuries of time and leisure and that their opportunities were no longer limited to the people who could reach them in person or through the mails. But where it was available, the line-of-sight telegraph was already capable of performing much of the feat — it was said that a message could reach Paris from the border in three minutes, though only by daylight and in clear weather — and in the great cities of Europe and America, the postal services were vastly superior to the postal services now prevailing. In London and New York, it was possible to send an invitation in the morning and find the recipient seated at table the same evening. Still, there was no magic in them. The line-of-sight telegraph and the swift postal services worked the way things had always worked. The telegraph did not. In the early days, it inspired in its observers something that resembled a fit of ecstasy, a rage, and a kind of postmodern suspicion.

One editorial writer, whipping himself into a froth, proclaimed that the electromagnetic telegraph was an event "next only in importance to the Crucifixion," while another was moved to write: "The railroad car, the steamboat, and the magnetic telegraph, have made a neighborhood of the widely dissevered states." The latter was correct, but in a way he had not foreseen. The first great use of telegraph was to direct the Union armies in the slaughter called the Civil War. The side that won the war was the side that made the best use of the telegraph. The neighborliness of the great colonial empires was similarly improved. A mutinous Indian sepoy, being led to his execution, was able to point to the telegraph line and exclaim, "There is the string that strangled us." New York City, as always, marched to its own drummer. Only after Benjamin Silliman assured the citizens that their health would not be affected was the telegraph allowed to enter the city.

* * *

The press — already converting itself in the major cities into a daily chronicle of rapine, trivia, and accidents — was able to flood itself with information that, some critics believed, was of absolutely no use to its readers. Previously, news from afar had been published weeks late, often inaccurately, and surrounded by the opinions, some of them hysterical, of the paper's owner. Now, with information passing in real time, the profession of reporter was born. The reporter was the man on the spot, recording what he heard and saw and sending it to his editor by telegraph. He was something entirely new. The reporter transformed the political life of the country.

But while some of the news was more interesting and valuable, because the newspapers now had it firsthand rather than second-hand and previously might not have heard of it at all, much of the news that began to appear was by all previous standards sensational and worthless. In the middle of the century a physician named George Miller Beard contemplated his fellow Americans and concluded that their minds were suffering from what we would now call sensory overload. Daily at the breakfast table, they filled themselves with irrelevant information of no conceivable value to themselves and jammed their intellectual circuits. The doctor put a name to the malady that resulted. He called it "nervousness." The modern profession of psychiatry was being born.

The vast modern industry of electricity-based devices was not.

The transformation of the telegraph into a citywide fire alarm system showed that conceptual leaps were now possible, but the demonstration is clear only in hindsight; it was not clear then. Although the Electric Revolution, like the parallel Industrial Revolution, was moving faster than anything the world had ever seen, people still thought in the way they had always thought, in terms of years passing through their own lifetimes, and sometimes decades passed before the next electrical advance was made. Moreover, the Industrial Revolution had not taught them to think differently about the process of invention. You made an invention, improved it, and then stopped. Then you made another invention. New inventions did not flow from old ones. It was not yet apparent that electrical invention did not work like that. For years after Morse, the progress of the electrical arts would proceed by the process that Stephen Jay Gould and Niles Eldredge called punctuated equilibrium: there would be a leap forward and then a stop, a leap forward and then a stop. It was not yet a revolution; for that, the world would have to wait for Edison and 1879. But it was possible, occasionally, to do some remarkable things.

* * *

By 1867 the loggers had cut their corridor through the forest and reached the Alaska frontier; they called it "the telegraph road." In their wake, work parties built huge bonfires to melt the permafrost and sink the poles. It was only a matter of time before the wires reached Nome, and the world telegraph system would become a reality. Everyone knew you could sink a telegraph cable under a body of water as narrow as the Bering Strait; by 1867 the British had done it often enough. Then orders came to suspend operations.

After years of trying, one huge false start, and an enforced delay while Generals William Tecumseh Sherman and Ulysses Simpson Grant battered the Confederacy into submission, a telegraph line from Ireland had reached Trinity Bay in Newfoundland in the middle of the previous year. For the first time, the continents were permanently joined.

CHAPTER FIVE

A RAILROAD TO THE MOON

Pay it out, Oh,
Pay it out as long as you are able;
For if you put the darned brakes on
Pop goes the cable.

— *Sung by the sailors of* Niagara *and* Agamemnon
to the tune of "Pop Goes the Weasel"

PART ONE: THE NEWFOUNDLAND PHASE

I n 1853, after he got out of jail, Frederic Gisborne returned to New York and began to look for money again. For one thing, unless he found some, he was going back to jail. For another, like Davenport and Morse before him, he was a man driven by a curious obsession. He was determined to build a telegraph line across Newfoundland, a place where almost no one lived.

He had been born in England, was partly raised in Tahiti and Mexico, and eventually settled in Canada. There, like a lot of likely young men in the 1840s, he found himself drawn to the new miracle of telegraphy, showed a talent for handling the new technology, and eventually rose to become superintendent of operations for the local telegraph company in Nova Scotia. He also came to believe that his fortune would be made in the Newfoundland telegraph on the Canadian island of that name, together with some carrier pigeons to relay the messages to the mainland, and eventually a short length of underwater cable. When all of these elements were combined, he anticipated, he would own one the most lucrative news operations on the continent.

* * *

Gisborne's thinking was more advanced than the technology of the day, or at least the technology that existed on the Canadian-American side of the Atlantic. In 1845 Morse had strung a line in New York harbor, proving that underwater telegraphy was possible over a limited distance, but he was unable to make a public demonstration of it because a passing vessel cut the line, and when he suggested that Europe and America could thus be linked, Congress greeted the statement with shrill hoots of derision. But though the conservative Americans weren't interested in undersea telegraphy, the British definitely were. They lived on an island and they had far-flung commercial interests, and an underwater electromagnetic telegraph was the obvious way of receiving news from the Continent in less than a day. The Britain of the mid-nineteenth century was a place that modern eyes would have found almost unrecognizable. It was a boldly entrepreneurial country besotted with new technologies. Its engineers and businessmen led the world, and they were willing to try anything. And they were the only people in the world who knew how to lay a commercially viable undersea telegraph cable.

The first telegraph cable reached France from England in 1851. By the following year, England was connected with Belgium and Denmark and trans-Mediterranean links were under discussion. By the middle of the decade there was a Black Sea cable linking Varna in Bulgaria with Her Majesty's forces fighting the Russians in the Crimea. This caused great consternation among the commanders in the field, who were then micromanaged by the politicians at home. Despite these achievements, no one knew how much electrical power to use, there wasn't much to begin with because all the man-made electricity in the world was provided by batteries, and everyone was employing variants of the landline telegraph equipment

Wheatstone and Vail had invented, making transmission times agonizingly slow.

Gisborne, although he later claimed otherwise, was hardly a visionary reasoning in advance of his times. He proposed to run a Newfoundland line from the colonial capital in St. John's, the North American city nearest to Europe and located on the Atlantic coast, to Cape Ray, the point on the island nearest to the Canadian mainland, link Cape Ray with the mainland by means of carrier pigeons (a stopgap: eventually he planned to lay a shallow underwater cable), and hook up with the Canadian telegraph system. Once complete, the beauty of his scheme would come into play: the Canadian telegraph system was connected to the American telegraph system. According to his plan, ships from Europe would stop at St. John's, where they would offload the latest news. The news would then be flashed to New York, where it would arrive two days earlier than usual, at which point Gisborne would begin to see profit, or so he hoped. News from Europe was a hot commodity in New York. Eager to be first to hit the streets with the latest from the Old World, local press barons dispatched fast sailboats to intercept ships as they entered the harbor, and reporters routinely bribed the captains. Clearly the newspapers in New York would pay a premium for news that arrived days, rather than hours, early. But, seamless and infallible though Gisborne's plan might have seemed, it was like many other plans: for it to succeed, other people with their own agendas were required to behave, rather precisely, in accordance with the plan. European ships would have to be persuaded to stop at St. John's, something they didn't normally do unless they were sinking. In modern parlance, St. John's was the back of beyond, and almost nobody wanted to go there. There was, in short, a serious flaw in Gisborne's thinking. It didn't stop him. He was a man of determination that, in other circumstances, would have been admirable.

The interior of Newfoundland was a place the size of England where no one, native or white, had ever lived. There were bogs that swallowed horse and man alive, unclimbed mountains, and uncharted valleys. In 1851 Gisborne surveyed his telegraph route with

Indian guides who appear to have been as clueless as himself, raised money in New York, set to work, and lost his shirt. After his New York backers failed to produce a second round of financing, there was so little money left that his work party began to starve; one of them actually did, and another was never the same again. Gisborne ended up in debtor's prison. One of the purposes of debtor's prison was to improve a man's thinking until he remembered where his money was, but Gisborne had sold all of his property in an attempt to cover his crushing obligations. Quite reasonably, he pointed out to his jailers that he wasn't generating any new assets as long as he was in stir. Quite reasonably, his jailers permitted him to return to New York and seek a fresh infusion of capital. It was New Yorkers, after all, who had given him seed money to build a telegraph line in Newfoundland. Anything seemed possible down there.

And it was at New York's luxurious Astor House, a rather curious venue for a man who claimed to be penniless, that by a happy chance Gisborne fell in with Matthew Field, a self-taught civil engineer who happened to have a brother who was looking around for some excitement.

* * *

Once, Cyrus Field was such a famous American that his deeds inspired the wretchedly bad verse that an earlier age lavished on its heroes. Field traced his ancestry back to John Field, the astronomer who introduced the Copernican theory to Elizabethan England, and like Morse he was the son of a New England divine. Field's eldest brother, Dudley, who lived next to him on Gramercy Park, was a prominent attorney who became Jay Gould's lawyer. Another brother, Stephen, was chief justice of California and, later, a federal Supreme Court justice who owed his appointment to his knowledge of Western business and property law. Cyrus, the seventh brother, was born in 1819 and thus became the first major figure in the development of electricity to be a child of the nineteenth century. Memorably described as slight and nervous, he was a man with a great gift for friendship. He was also a man with a great deal of money. The paper business had been extremely good to him.

In the 1840s the American paper business resembled the cable television business of the 1950s. The barriers to entry were very low, no special skills were required, and a great deal of money could be made. A prospective titan of the trade would obtain — cheaply — a plot of land in the Housatonic Valley of western Massachusetts, erect a mill at no great expense, and wait for the cash to roll in. Cyrus Field had once dreamed of just such a life. He trained in the mill of his brother Matthew, the engineer, and bought a mill site of his own. Then something occurred to him. He took an office in New York City — took one over, in fact, when it fell into bankruptcy — and became a wholesaler to the paper industry. His fortune was made. He had a gift for sniffing out popular trends, and it served him well.

He sensed when a market was developing for colored paper, alerted his customers, and sold them the raw materials. The same thing occurred with coated paper and paper made from European and Turkish rags. European rags were no different from any other rags, but to the American public Europe was the place where all good things came from. Cyrus Field, having timed the opportunity to perfection, happily supplied the necessary rags. He bought his splendid home on Gramercy Park, which was then on the suburban fringe of the city, grazed his cow in Madison Square, and collected the paintings of the Hudson River School. His home was the first in New York to be decorated by a professional. Cyrus Field always liked to lead the pack.

By 1853 his work no longer absorbed him. True, when he was in town he found it hard to stay away from his office. The name over the door was his own and he was the presiding genius, but his relationships with his customers were established, durable, and no longer absorbed his time. His lackluster brother-in-law could handle the day-to-day affairs that required no special inspiration. Cyrus Field was not exactly retired — he would, in fact, never retire, although he eventually put the paper trade behind him — but he could take time off. He had recently done just that, to tour South America, where the only words of Spanish he learned were *más rápido*. He was only thirty-three, and he liked things done swiftly.

And so it was to a man who could take time off, a restless man whose wholesale business had matured and whose early challenges were behind him, that brother Matthew brought Frederic Gisborne and his Newfoundland dream. Because Field left no memoirs, his inner life is something of a mystery, and his biographers do not record his impressions of his brother's new companion. Later comments suggest that Field was not overly taken with the man, who soon vanished from his life. But Field was polite enough to give Gisborne a hearing, and once his visitors were gone, he began to pace his study. In his study was a globe of the world.

It was a time when people pored thoughtfully over the newly accurate maps, and Field marshaled his thoughts while contemplating his globe. All he knew about telegraphy came from the newspapers, and all he knew about geography — the recent South American trip excepted — was under his nose. There, he noticed something that had not occurred to him before, in part because he hadn't been certain where Newfoundland was. His globe informed him that the shortest distance between America and Europe was the distance between Newfoundland and Ireland.

Cyrus Field resembled Morse in one important way. When it came to the telegraph, they both began as ignoramuses. Indeed, Morse was one of the most successful ignoramuses in world history. Although he dithered for years with the technological breakthrough that had come to him in a single moment of insight, he worked at it stubbornly until someone taught him how to do it correctly. Field in his study, like Morse on the *Sully*, had also been seized by an insight, but he differed from Morse in one important respect: he knew what he didn't know. For one thing, he had no idea what lay beneath the blue surface of the Atlantic. Only one person in America did, and in a distant way, Cyrus Field was acquainted with him. He was Lieutenant Matthew Fontaine Maury, the superintendent of the Naval Observatory in Washington, and Field had been quite taken by one of his books.

Memoirs of Maury usually begin with the comment that he was the greatest scientist the country has ever forgotten. Farm-born in

Virginia, Maury entered the navy as a midshipman in 1825 at the age of nineteen. Nine years later, a service that did not spoil its officers with promotion finally made him a lieutenant. Eight years after that, he took charge of the navy's observatory. In the interval, he had created the science of oceanography.

Until Maury, most maps of the world were splendid guides to the landmasses that surrounded the world's oceans, but on the oceans themselves, they were silent. As a result, ship's captains had an excellent idea of where they wanted to go, but they navigated the intervening waters by memory, guesswork, and astronomical calculations. Maury changed all that.

In his charts, papers, and books, he described in painstaking detail just what a captain could expect in the way of wind, currents, tides, and weather during the course of his voyage. It was a monumental achievement and an extremely useful one. During the Gold Rush, using Maury's instructions, the average time it took a clipper ship, the fastest commercial vessel on earth, to travel from New York to San Francisco fell from 180 days to 133. No one knew more about the oceans than Maury did, and Field thought he might be able to help.

Although he was extremely careful when it came to describing the ocean, like many sea dogs Maury seemed to take leave of his senses when he was set loose on land. In Maury's articles on South America in the *National Intelligencer,* Field had read of how gold could be sifted from the streets and how diamonds were to be had for the plucking. The wanderlust was on Field then, his purse was long — his personal fortune was somewhere in the vicinity of $250,000, an enormous sum — and he convinced himself that he had two pressing reasons to travel in Latin lands. Some years earlier, his brother Timothy, the family's only sailor, had disappeared on a voyage out of New Orleans. The family was certain he was lost at sea, but a cruel rumor had reached their ears that Timothy had married and grown wealthy in South America; as a result, no monument had been erected to him in the family plot. Guided by Maury, Cyrus could now discover the truth. And, following Maury's leads to the untold riches of the southern continent, Field believed he could undoubtedly establish some commercial venture there that would be the making of his second fortune. He enlisted the aid of another paper magnate, who never showed up at their planned South American rendezvous, and he recruited the paper heir and Hudson River artist Frederic Edwin Church. Church spoke Spanish, and his paintings — including the paintings Field later hung in his home — would serve as advertisements for his unknown new venture.

Field found neither his brother nor his second fortune. He returned to New York with a wealth of experience, a flock of parakeets, a jaguar on a leash, a grass suit that he wore in the street to the delighted whoops of the onlookers, and a native boy who took to lurking in the darkened hallways of the Gramercy Park house with a knife, frightening the servants and the family until Cyrus went on another trip and the family sent him home. Then and all his life, Field was an impulsive, romantic, slightly naïve, humorous, and sometimes slightly preposterous man. But despite the fact that South America had fallen somewhat short of Maury's descriptions, Field knew that there was one area in which Maury was unsurpassed. If anyone knew what was on the bottom of the North Atlantic, it would be Matthew Maury. Field immediately sat down and wrote him at the observatory.

"Curiously enough," Maury replied, "when your letter came I was looking over my letter to the Secretary of the Navy on that very subject." Maury, too, wanted to know what was on the bottom of the North Atlantic, and he had dispatched Lieutenant D. H. Berryman in the navy brig *Dolphin* to find out. Berryman was equipped with the latest wonder of the age, called Brooke's Apparatus. It was a cannonball threaded with a tube of metal fitted with hinged jaws, and it took an hour to descend through a mile of ocean water. When it reached bottom, wires leading to the surface moved a swivel that released the cannonball and closed the jaws of a kind of posthole digger at the lower end of the tube, trapping a sample of the bottom sediments. The tube was then pulled back to the ship.

The modern equivalent is a machine for taking a core sample. "The result is highly interesting," Maury reported to the secretary of the navy.

> From Newfoundland to Ireland, the distance between the nearest points is about 1,600 miles and the bottom of the sea between the two places is a plateau, which seems to have been placed there especially for the purpose of holding the wires of a Submarine Telegraph, and of keeping them out of harm's way. It is neither too deep nor too shallow; yet it is so deep that the wires, once landed, will remain for ever beyond the reach of vessels' anchors, icebergs, and drifts of any kind, and so shallow that the wires may be readily lodged upon the bottom. The depth of this plateau is quite regular, gradually increasing from the shores of Newfoundland to the depth of from 1,500 to 2,000 fathoms as you approach the other side.

Moreover, the bottom was paved with an even layer of microscopic shells, indicating that subocean currents did not disturb it, either. "I [do not] pretend," Maury continued,

> to consider the question as to the possibility of finding a time calm enough, the sea smooth enough, a wire long enough, a ship big enough, to lay a coil of wire 1,600 miles in length; though I have no fear but that the enterprise and ingenuity of the age, whenever called upon with these problems, will be ready with a satisfactory and practical solution of them.[1]

To Field, this was a huge stroke of luck. The bottom of the Atlantic between Ireland and Newfoundland was a nearly level plain and Field, reading Maury's words, realized that the Telegraph Plateau, as he came to call it, could have been designed for his purposes. Remembering that Morse had predicted an Atlantic cable, he sent a letter to the inventor. Morse responded by coming in person and soon consented to join Field's proposed venture with the title of "advising electrician." In his correspondence, however, Morse described himself as the venture's chief electrician. He also had no idea what would happen if a telegraph cable were dropped to the bottom of the Atlantic, or whether a battery could power it. No one did. All the world's underwater cables had been laid by the British; no American, with the exception of Morse's one failed attempt in New York harbor and a successful attempt in the Hudson River by Ezra Cornell, had any experience with them. And all the underwater cables were laid in shallow seas. The Atlantic was not shallow. What would great depth do to a telegraph cable? Nobody knew that, either.

There were many other unknowns. For example, there was a school of thought that at some point in the depths the ocean's pressures would equalize themselves — the pressure from below counteracting the pressure from above — and beyond this point a telegraph cable could not sink. To Field, this was a small matter. He believed that if Brooke's Apparatus — the sampling device that Maury's people had used that looked like a combination of a cannonball and a scoop shovel on a line — had made it to the bottom of the Atlantic, a heavy cable would, too. As for the cable itself, the British would provide it. They were the only people on earth who knew how to make one, and they were in the business to make money. Indeed, the principal undersea cable makers in Britain were pit rope manufacturers, companies that had previously specialized in making the hoisting ropes for coal mines, and they had branched out in search of a new opportunity.

At first Field thought that ten wealthy New Yorkers would be enough to provide the necessary money, starting with his Gramercy Park neighbor, the sixty-two-year-old glue millionaire and philanthropist Peter Cooper. Cooper, too, had his eye on the main chance. "If ten men can carry out the project," he said, "so can five." Five men it was, then, Field and Cooper and three others. The fewer of them there were, the larger the shares of the inevitable profits.

In stark contrast to Morse, Field and Cooper were both seasoned men of business who knew exactly how to set about their task. First, they hired Field's brother Dudley as their lawyer. Next, they paid off Gisborne's debts and secured the rights granted him by the Newfoundland colonial government to construct a telegraph line. Last, Field went to England to buy a quantity of marine telegraph cable to complete the link from the island to the mainland. This was the first of more than forty transatlantic trips he made in pursuit of his telegraph venture. When it came to handling governments and closing deals, he and Cooper were in their element. However, when it came to laying a transatlantic telegraph line, they were in territory that was entirely new to them — and to everyone else. The previous record length of undersea cable was 110 miles. The distance from western Ireland — the closest point in the British Isles — to Newfoundland was 1,900 miles. The previous record depth was three hundred fathoms — about 1,800 feet. The Atlantic was thousands of fathoms deep.

In London, Field had no trouble in obtaining the necessary cable for the short leap from Cape Ray to the mainland, an experienced English cable engineer, and a vessel to carry them both to a meeting place in Newfoundland. Then he returned to America to complete the promotional part of his plan. He and the four other "projectors," as they were called, were going to need a spot of help from Washington, and it was important to alert and prepare the governmental mind by putting on a little show.

A comfortable vessel was chartered from the Carolina Steamship Company and invitations were sent. Morse would lecture on his telegraph, and there would be a band and all the entertainments of a recreational cruise. Prominent members of the clergy were asked aboard, and so were prominent men of New York and their ladies. Amid music and a fluttering of handkerchiefs, the assembled party gaily set forth to Newfoundland, where its members would witness the laying of the mainland cable. The food and drink were of the best.

It was at this point that the undertaking began to descend into low comedy. The problems began when one of the prominent clergymen occupied the captain's chair during lunch. The captain, who appears to have been a remarkably difficult man, was not inclined to forget such a slight to his dignity, rank, and uniform. When the steamer met the sailing vessel carrying the cable and engineer from England in Newfoundland waters, everything began to go wrong. Because they knew nothing about laying cable from a ship, they made a beginner's mistake: they tried to lay the cable from a sailing ship. Then the Carolina captain's ego got into the act.

It was decided to take the English vessel under tow. The latter would then lay the cable under the engineer's supervision. First, performing what appeared to be a simple maneuver, the Carolina captain rammed the cable ship. Then he ran it onto a reef. Then, matters having been corrected, and with the cable playing out nicely behind the English vessel, the Carolina captain set off in the wrong direction. Peter Cooper ascended to the bridge of the ship, reminded the captain that their destination in Nova Scotia was marked by a rather large flag, and ordered the captain to steer for it. The captain, outraged that a civilian who had paid for his ship had the temerity to indicate its destination, replied, "I know how to steer my ship. I steer by my compass." He continued to head in the wrong direction. Cooper had brought his lawyer along for the trip. This worthy drew up a formidable legal document, Cooper presented it to the captain, and the captain began to zigzag; over the course of the next nine miles, the towed English cable ship played out twenty-four miles of cable. The captain also increased his speed, the cable-laying apparatus could not keep up, and the smaller cable vessel was in danger of being pulled underwater. The English engineer cut the cable and saved the day.

At this point, a British patrol vessel appeared, sporting an enormous sign: DO YOU NEED ASSISTANCE? The Carolina captain did not reply. The patrol ship raised another huge sign: ANSWER YES OR NO.

It also seemed to mean business. Field and Cooper found brushes and a can of paint and wrote NO on the side of the funnel. There would be no further cable laying; there was no longer enough to reach the mainland. On the voyage back to New York, the distinguished clergymen and other guests held a ball in honor of Field and Cooper and declared that they'd had a wonderful time.

A lesson had been learned: sailing vessels were not suitable for the job, even when they weren't being towed by a madman. Only steamships would be used. This presented the projectors with another problem. Nowhere in the world, as far as they knew, was there a steamship capable of carrying two thousand miles of heavy cable. Two or more large vessels would be needed — which presented yet another problem. The projectors were running out of money.

The wire for the onshore Newfoundland line to bisect the island from St. John's to Cape Ray was provided by one of Cooper's companies, and the engineer Matthew Field was placed in charge of the six hundred men who would build it in accordance with the contract from the colonial government, which required the construction of a road eight feet wide equipped with bridges of the same width. The work was scheduled to take a year. It took two, and it cost a million dollars. The projectors had raised only a million and a half, and $350,000 of it had disappeared over the side when the first underwater cable was cut. Clearly, another trip to England was in order. This time, Field was accompanied by Morse.

PART TWO: THE FIRST TRANSATLANTIC ATTEMPT

Field was gradually learning his new job. Although he and Morse were very different, they had one thing in common. In the first flush of enthusiasm — Morse for the telegraph, Field for the Atlantic cable — they had seriously underestimated the immensity of the task before them. Morse, the artist, had solved his problems by accidentally falling in with the right crowd. Field, the seasoned businessman, set out to recruit a "right crowd" of his own. England was still the only place where he could obtain gutta percha and the necessary cable expertise, but he had learned much from the farce off the coast of Newfoundland. To get some idea of what lay before him, he would go directly to the top: Faraday, who was in England. Apparently Field had never heard of Joseph Henry, now the founding secretary of the Smithsonian, and Morse was in no mood to enlighten him. Britain was also the home of William Thomson, a brilliant young Scottish professor of physics who had recently written about the possibility of an Atlantic telegraph cable and included some fairly advanced-looking calculations. And there was Edward Orange Wildman Whitehouse, a medical doctor who had made something of a name for himself as an electrical experimenter. Field was impressed by credentials, which was why he was so enamored of Morse. Faraday, fortunately, was as brilliant as ever, but Field had no idea that Whitehouse's theories of undersea telegraphy were based more on fantasy than science. He also had no idea that Thomson was the luckiest choice he would ever make in his life.

William Thomson's own biographer said that during the course of his long life — he was born in 1824, just in time for the first railroad train in 1825, and died in 1907, four years after the Wright brothers achieved heavier-than-air flight at Kitty Hawk — he was a man incapable of doing anything wrong in the first half of his career and incapable of doing anything right in the second half. Motherless from the age of six, Thomson had written his first important paper at the age of fifteen and became a professor of physics at Glasgow when he was twenty. Like Benjamin Thompson, who became Count von Rumford, the Scot William Thomson was an expert on heat theory, and he completed Rumford's demolition of phlogiston and caloric — two supposedly invisible and weightless substances that supposedly caused combustion, had dominated scientific thinking on the subject for over a century, and did not exist. William Thomson concluded that heat and motion were identical, and he charted the similarities between heat and electricity. Still, despite his paper on a possible Atlantic cable, it was a bit of a stretch for Field to invite him aboard the actual venture. Thomson was not a forceful man, and he was a professor of thermodynamics with a marginal interest in the electrical phenomenon. But he was an inspired choice.

Field also needed money and a very large ship. On his previous trip, he had made the acquaintance of John Brett, the former antiques dealer who laid the first undersea cable to France in 1851 with the help of his brother Jacob. The secret of their success was insulation made of gutta percha, the gum of the gutta percha tree that served the Victorians as a kind of plastic. Dolls were made of it, and so were overshoes. Wheatstone had suggested it: gutta percha, unlike rubber, didn't lose its flexibility when it dried, and thanks to the British colony in Malaya — the only place on earth where the gutta percha tree grew — they had a monopoly on it. The Bretts learned as they went along. For example, they learned that if their cable was too heavy, it played out too fast from the stern of their boat and their supply was soon exhausted. They strung their line across the Channel and hooked into the Continental telegraph network, but not for long. A French fisherman hoisted their cable to the deck of his vessel, cut into it, saw the gleam of copper, believed he'd discovered a form of gold-bearing marine life, and took a piece home. The Bretts had to lay a new cable. But their pioneering work had shown the way. Now John Brett was the head of England's Magnetic Telegraph Company. Of the $1,700,000 in shares floated to revive Field's venture, Brett's shareholders bought most of them. In fact, from this moment onward, the Americans began to lose control of the project. In the future, most of Field's money would be raised in Britain.

As Britons, Brett's shareholders were well aware of the unique commercial opportunities involved. They were also, it can be presumed, somewhat better informed on the subject than the vast body of their countrymen. When Brett ran his first wire under the Channel, a number of people thought it operated like a bellpull, that a jerk on the English end would show up as a jerk on the French one, spelling out a message, and the state of telegraphic knowledge had not much improved in the time since. When word leaked out of Field's venture, one man offered to lay the entire cable himself, if Field would buy him a rowboat and a marlinspike. Others, only slightly less impractical, suggested supporting the Atlantic cable with buoys or, in a way that was never made very clear, parachutes; ships expecting or needing to send a message could sail up and hook on. Still others, including the astronomer royal, G. B. Airy, pronounced the whole idea preposterous. Field, they said, was planning to build a railroad to the moon. The British government did not agree. The British government, like Brett's shareholders, saw promise in the venture.

Field, having grown cautious with recent experience, asked the British government for a new survey of the Atlantic seabottom. It would be the third such survey. Maury, at Field's request, had dispatched a second American vessel as a check on the findings of the first, and it confirmed them: the bottom of the North Atlantic, just as Maury had said, was a submerged plain ideal for a telegraph cable. A British vessel was swiftly assigned to the task, again confirming Maury's findings. Field requested a naval vessel to lay his cable, and one was promised. And the British government offered a handsome subsidy for the cable, an annual sum of $70,000. Not only were the investments of British subjects at risk, but Britain was bidding fair to become the center of a new global communications system, which it eventually did. And the British government would, of course, expect priority to be given to its own messages.

Morse and Whitehouse, through the good offices of Brett, successfully sent a telegraph signal through a thousand miles of the Magnetic Telegraph Company's landlines and pronounced the certain success of the Atlantic cable. It didn't occur to either of them that sending a signal through a submerged cable was a vastly different undertaking than sending one through a landline. Certain problems did, however, occur to the wiser heads of British science and engineering. Faraday, backstopped by the much younger Thomson, predicted that an induction field would form around the wire. There was also the possibility that the cable would begin to act as a capacitor — that the metal-mesh shielding that protected it would store a certain part of the electrical charge. This would create an effect called inertial induction, which would diminish the current in the core of the cable itself, and the possibility should be taken into account when designing the sending equipment. Ohm's Law, which said the flow of electricity through a conductor depended on the amount of resistance the electricity encountered, would again come into play. Whitehouse, who did not understand Ohm's Law, found the possibility of this happening remote. He proposed to drive an enormous battery current, the largest he could generate with the help of electricity-enhancing electrical coils, down the cable. He would thus, he thought, overcome any such obstacles as resistance by brute force. He did not understand that in the cable it would be the voltage of the current and not the amount of the current that did the job. The cable would function like a garden hose. The water coming out of the end of the hose was analogous to the current, which is called amperage. (When you are electrocuted, it is the amperage that kills you.) But if you covered half of the hose's end with your thumb, the stream of water would suddenly intensify, although there was no more water in the hose than formerly. This was how increased voltage worked. Professor Thomson, as shy and diffident as Whitehouse was aggressive, understood this very well. It led to his greatest invention, the mirror galvanometer. Whitehouse's failure to understand it led to his complete and utter failure.

There would be, Faraday explained, a slight delay in the transmission between London and New York. He thought it would be about a second. Although Thomson and Charles Bright, the twenty-four-year-old English technician who was selected to be the transatlantic venture's chief engineer, believed the cable should be rather thick, Faraday insisted that it should be less than an inch in diameter. It's hard to know what Morse thought of all this, but he forcefully seconded Faraday. Faraday and Morse were wrong; experience subsequently proved that a thicker cable was the way to go. Ever afterward, for as long as copper was the vehicle of undersea transmission, Thomson and Bright's calculations were used.

Field was completely out of his element in these discussions, but he came into his own again when it was time to manufacture the actual cable; manufacturing, like negotiating with governments, was something he understood. The core of the cable would be made by the Gutta Percha Company of London. For the first time in the history of electric transmission, it would be a redundant system rather than (as with Morse's line between Washington and Baltimore) a single wire. The cable would have seven wires surrounded by insulation, and each wire would be capable of handing the cable's entire traffic of messages. Thomson said the copper should be of unusual purity. Whitehouse said it didn't matter, and Whitehouse carried the day. Cyrus Field, flush with British money, was in one of his hurries and following Thomson's suggestion would take time. To save even more time, Field assigned the armoring of the cable — the woven metal cage that would surround and protect the cable's vital core — to two different companies, Glass, Elliott of London and R. S. Newall of Birkenhead, former pit rope manufacturers. Later, there were those who questioned the wisdom of assigning the work to separate companies, neither of which knew what the other was doing. Perhaps Field had gone a bit too fast, and, indeed, when the separate halves of the cable were delivered, one was armored on a right-hand spiral and the armor on the other had a left-hand spiral. It was a minor thing. The operation of the cable was not affected. But what other corners had been cut?

They were flying blind. The Atlantic cable was an entirely new thing — there was no useful precedent for it in shallow-sea cables — where it seemed that one man's guess was as good as another's. Only in retrospect would it be realized that this was not so. For example, gutta percha was a substance that should have been stored under water, as it deteriorated in direct sunlight. Later and wiser, Field's people would go to great lengths to protect their insulation, but at some point during the handling of the first transatlantic cable, it was stored dry in the open. It was the wrong diameter. Its copper wasn't tested for purity. At the time, none of these things seemed to matter. Repeated tests showed that a signal flowed through the cable's entire more-than-thousand-mile length, which was the purpose of the thing. But they would never manufacture and handle a cable so carelessly again.

While the cable was being manufactured, Field returned to America, spent Christmas with his family, and went on to Washington. His experience in Britain, which he could now use as an example, led him to believe that he would have little trouble raising yet more money in his own country. Although the project enjoyed the support of Secretary of State William Seward and President Franklin Pierce, Field should have known better, but he hadn't stopped to research the matter. Only a handful of Americans wanted to buy the shares, and Field's considerable resources of charm and persuasiveness could not budge the others. American capital, wrote John Mullaly of the *New York Herald*, "was needed for the execution of the enterprise which the confidence of moneyed men in the United States did not enduce them to supply."[2] In the end, Field bought most of the shares himself. For better or for worse, the bulk of his fortune was now tied up in the Atlantic cable.

But to actually lay the cable, he needed what the British had already given him: an annual subsidy. He also needed a large ship. The Southerners in Congress weren't inclined to give any kind of ship to him. Their arguments were both accurate and utterly wrongheaded. They pointed out, rightly, that the cable was now largely a British project. Virtually all the new money was British. The cable would start in Ireland, a British possession, and terminate in Newfoundland, a British colony. Senator Stephen A. Douglas of Illinois, the leader of the pro-cable forces, pointed out, also rightly, that neither of these things mattered. The benefits of the cable would redound equally to the British and the Americans. The commercial advantages were huge. The political advantages were obvious: never again would Americans and British kill each other after peace was declared or declare war on each other after a *casus belli* had vanished. If war did come between the two countries — a point the

Southerners had raised and emphasized — the British would inevitably seize control of the cable no matter who paid for it, and if the British wanted to lay a cable on their own, freezing out the Americans entirely, they were perfectly capable of doing so.

The advantages of Field's plan were obvious. Citizens of both countries would own the cable, and the two countries would have equal access to it. The British, without being asked, had already ceded the point. The Southerners were unmoved. Wrapping themselves in the mantle of patriotism, they vowed to defend to the last the hard-pressed American taxpayer against Albion's wiles; two of their senators, James M. Mason and John Slidell, were particularly strong on the subject. Yet the subject, when closely examined, had remarkably little to do with what was being talked about. The steamboat, the railroad, and the telegraph arrived on the scene at roughly the same moment, between 1810 and 1840, a blink in historical time, but the British had had seventy-five years to accustom themselves to the lightning pace of change in the Industrial Revolution. The Southern legislators, like the American investors who refused to part with their money, had not. They had greeted Morse's telegraph in much the same way. They were suffering from future shock.

In the end, the subsidy bill passed the House in 1856 by nineteen votes and the Senate by one. As one of his last acts in office, President Pierce signed the bill into law. It was immediately challenged as unconstitutional. The attorney general ruled that it was not.

And Field was provided with a ship — a ship, in fact, that more or less made up for his trials in the halls of Congress. The brand-new USS *Niagara* was rated as a steam frigate, but she was the largest warship in the world, at 5,200 tons some 2,000 tons heavier than the largest battleship in the Royal Navy. She was also proof positive that the Americans, when they chose to stretch themselves, were no slouches at technological innovation. *Niagara* had been designed by George Steers, the builder of the magnificent yacht that had brought the America's Cup to New York. She was flush-decked, and her planned armament, not yet aboard, consisted of twelve of the gigantic new Dahlgren guns, each served by a crew of twenty-five and capable of throwing a 130-pound projectile for a distance of three miles. Her single-screw propulsion system gave her a rated speed of between ten and twelve knots. She was, in all, the most modern warship in the world and one of the most magnificent, and from Field's point of view, she possessed an incomparable advantage: she was very, very large. He could get a lot of cable into *Niagara*.

With Field and Morse embarked, she sailed for England on April 24, 1857. Two observers from the Russian navy were present but the press was not invited aboard, although Morse, always with his eye cocked in the direction of fame, had hired Mullaly of the *New York Herald* as his private secretary. In the Thames estuary they were met by *Agamemnon*, the principal British contribution to the telegraph fleet. She had been the flagship during the bombardment of Sebastopol during the late Crimean War, and although she was equipped with steam, she was otherwise a vessel where any veteran of Trafalgar would have felt perfectly at home: an old-fashioned two-decker with a high ornate poop. But she had a distinguishing feature that recommended her for the task at hand: her engine room was located aft, giving her an unusually large hold. Even so, much of her cable cargo would have to be coiled on deck, as would tons of bagged coal. Like *Niagara*, she was fitted with a guard, known as a crinoline, that prevented her propeller from severing the cable as it was played out. She had been jury-rigged with a lighter top-hamper than usual, making her less maneuverable. The cable, when out of the water, weighed just under half a ton per mile. *Agamemnon* carried eight hundred of those miles. When loaded, she was dangerously down by the head and hard to steer. She leaked — her seams gaped by an inch, she lacked adequate sails, and her decks were cluttered with potential projectiles. In fair weather, with a light sea, none of this would be any problem.

When it came to the actual laying of the cable, there were two schools of thought. Thomson, Bright, and the other British engineers wanted to proceed to midocean, splice cables, and send the ships off in opposite directions, which — all other things being equal — would halve the amount of time involved, reducing *Agamemnon*'s exposure to any possible dirty weather. The electricians, led by Whitehouse, wanted to start at Ireland, with *Niagara* striking a blow for transatlantic amity by paying out her cable in British waters. The cable fleet, escorted by two frigates, one British and one American, and three smaller craft, one to set the course and two to bring the cable ashore at the far shore, would then proceed to midocean, splice, and give *Agamemnon* the honor of laying the rest of the cable to North America. This would have the advantage, not readily apparent to Bright and the engineers, of allowing the cable fleet to remain in telegraphic touch with the cable's terminus at Valentia, Ireland, the westernmost town in Europe. Whitehouse had decided that he was too ill to travel with the fleet. Professor Thomson, who had joined the venture's board of directors, would take his place on *Agamemnon*, although young Thomson had no official standing whatever. He wasn't even being paid.

Finally, however, there was something resembling popular enthusiasm for the project back home in America. "I hold in my hand," announced the popular orator Edward Everett at the opening of the Dudley Observatory in Albany,

a portion of the identical electrical cable, given to me by my friend Mr. Peabody, which is now in progress of manufacture to connect America with Europe. Does it seem all but incredible to you that intelligence should travel for two thousand miles, along those slender copper wires, far down in the all but fathomless Atlantic, never before penetrated by aught pertaining to humanity, save when some foundering vessel has plunged with her hapless company to the eternal silence and darkness of the abyss? Does it seem, I say, all but a miracle of art, that the thoughts of living men — the thoughts that we think up here on the earth's surface, in the cheerful light of day — about the markets and the exchanges, and the seasons, and the elections, and the treaties, and the wars, and all the fond nothings of daily life, should clothe themselves with elemental sparks, and shoot with fiery speed, in a moment, in the twinkling of an eye, from hemisphere to hemisphere, far down among the uncouth monsters that wallow in the nether seas, along the wreck-paved floor, through the oozy dungeons of the rayless deep?[3]

It was an enthusiasm of a very early-Victorian sort, of course, and Everett was not alone in evoking images of drowned and cascading bodies when he contemplated the rayless deep. But the telegraph, with its relentless demands for concision and brevity, was already putting paid to this sort of windy, clause-upon-clause, adjective-bedecked form of utterance. It was Everett who gave the hours-long warm-up speech at the dedication of the Gettysburg battlefield. He was followed by Abraham Lincoln, who with a few short words re-defined forever the American political speech. Along with his study in the White House, the War Department telegraph office was where Lincoln spent most of his war. It is possible to track, in the increasing brevity and simplicity of his speeches, his growing familiarity with the language of the telegraph.

In Ireland, the telegraph fleet was off at last, bid godspeed by yet more volumes of windy blather from the lord lieutenant himself, Lord Carlisle, who personally helped to wrestle the cable ashore. In part because His Lordship's speech took up so much time, *Niagara* and *Agamemnon* waited until the following morning, August 5, 1857, to set out on their way, with their ship-to-shore batteries packed in sawdust to keep the acid from spilling onto the deck. Each ship was equipped with a new electrical speedometer that seems to have resembled a submerged paddle wheel. As it revolved, it made and broke a battery-powered current which then recorded the revolutions on some sort of "apparatus" on deck. It may not have worked very well, as nothing more was heard of it.

Five miles out, the cable tangled in *Niagara*'s paying-out machinery and broke. Field and his team were new to this, too. No one had ever tried to lay a cable that weighed anywhere near as much —

261 pounds per nautical mile once it was in the water, whose buoyancy would slice nearly 700 pounds off its dry-land weight — but
even in the shallow Mediterranean, Brett had warned them, he had
once lost a heavy cable when his machinery couldn't hold it and the
entire length had unspooled from its drums and vanished over the
stern into the sea. Taking Brett's experience to heart, Field's team
had deliberately overengineered their own machinery. The cable-
laying "engine," as it was called, took up most of *Niagara's* fantail,
where it passed the cable over a multitude of grooved wheels. It was
equipped with powerful brakes. But overengineering meant complexity, and complexity meant that many things could go wrong.
One of them had. The ship returned to Valentia, underran the laid
cable, found the end, made a splice, and proceeded on her way at the
cautious speed of two knots. Even this was too much for Morse. He
did not, in the words of one chronicler, "withstand the journey well."
Constantly seasick, he remained in his cabin for most of the voyage,
and he never joined another cable-laying expedition again.

Slowly *Niagara's* speed was pushed up to the estimated optimum
of five knots. Continental telegraphic contact was maintained with
Whitehouse in Valentia. Aboard ship, a strange silence reigned,
broken only by the continual thrumming of the cable machinery's
multiple wheels. The crew seemed to regard the cable as though it
were alive, skittish, and liable to bolt at any moment. By the second
day, they had reached the two-hundred-mile mark. The depth of
the ocean was 1,750 fathoms, an increase of 1,200 fathoms in ten
miles. This was expected. Maury's two hydrographic expeditions,
and the Royal Navy's single one, had indicated a gradual falloff as
Ireland was left behind. Soon they would be laying cable in 2,000
fathoms of ocean. Field's team quickly discovered that increasing
depth improved the quality of the transmissions.

On the evening of August 10, five days into the trip, the cable
went dead. Morse, summoned from his cabin, could not revive it,
and neither could the British technicians. After two and a half
hours of silence, plans were made to cut the cable. Then, suddenly,

it came alive again. No one knew why. Morse, hazarding an explanation, revealed that he knew nothing about how a short circuit worked. Perhaps, he suggested, the insulation had somehow been stretched, but the pressure of the ocean deep had forced it back together again. While the other technicians were pondering Morse's words, the cable broke.

At the cable-laying machinery in the stern, an accident had been threatening to happen for days. Bright's engineers knew enough about the laws of physics to understand the trick of laying cable. As the stern rose, you played it out. As the stern fell, you slowed it down. You played the cable like a fish. Later Bright placed the blame for the accident on the fact that his people were exhausted by the weather. This seems disingenuous: there was nothing particularly unusual about the weather, and future expeditions did not suffer from it, at least not that way. But Bright and his people had designed the cable-laying machinery, they were deeply committed to the project, and they were answerable to the stockholders who had put up the money. It might not have been politic to reveal that the grooves in the machinery's wheels were too shallow, for the stockholders might wonder what else was wrongly designed and begin to worry about their money. The cable kept threatening to slip out, to tangle, to jam the mechanism — and a jammed mechanism would be a disaster, because there was no way to stop the ship before the cable parted. Somehow, Bright's engineers had kept the thing running, but by August 10, only one of them was still standing. And Bright, standing with the engineer in the stern, decided to take some time off.

There was never a clear explanation of what happened. Morse, for once, is the only relatively disinterested witness (although he did have $10,000 of his own money riding on the cable's success), but his secondhand version of events is about as muddled as everyone else's. This much is certain: for whatever reason — fatigue, stupidity, or a simple mistake — while Bright was absent from the stern, the brakes were applied to the cable on the upswing, the cable snapped, and the broken end was soon two miles underwater. There was nothing to do but turn around and go home.

There were no recriminations. Everybody understood a simple mechanical accident, ropes and cables had been breaking for thousands of years, there was too much money involved to turn back and not too much to abandon the whole thing as a bad job, and the cable-laying season was over. The unused cable was stored ashore and seven hundred new miles were ordered. The Americans undertook to redesign the machinery, and the British contributed an anti-lock brake. The new equipment was smaller, more compact, and reengineered in the light of experience. It occupied only a third of the deck space taken up by its predecessor, it weighed a quarter as much, and the cable passed over two wheels rather than four. Less complexity meant fewer things to go wrong. The new brake was controlled by a series of weights. With this new system, the breaking strain on the cable could be computed with fair accuracy, and it was now possible to keep the brake pressure under that critical point. Although the cable could be slowed, only the engineer in charge of the brake could stop it, because he was the only person who had access to the weights. A sailor standing at a wheel that resembled a small ship's helm would continually monitor and adjust the strain. Except under extraordinary circumstances, stopping the cable at sea was to be avoided at all costs.

PART THREE: THE SECOND TRANSATLANTIC ATTEMPT

In view of the eminence William Thomson shortly achieved, when he became Faraday's recognized successor as the most eminent scientist in Britain, with first a knighthood and then a patent of nobility and more initials after his name than anyone in the British Empire, it is odd to think that he played no active role in the actual laying of the cable. Bright was the chief engineer, Whitehouse was the chief electrician, and the electrician aboard *Niagara* was young Charles de Sauty. Thomson, not even rated as an employee and assigned to *Agamemnon*, was supercargo aboard a ship that remained a bystander. But beneath his diffident exterior, his mind was working furiously. He disagreed with Faraday about some things — the diameter of the cable, for example — but they were in accord about the possible retarding effect of a cable-generated induction field. Unlike Whitehouse, Thomson understood the implications of Ohm's Law. The brute force solution of pouring more current into the wire would not work and might lead to unpleasant consequences. The solution to undersea cable transmission, Thomson realized, was to reduce the current to a whisper. But whispered power wasn't strong enough to create a reaction in the receiving apparatus, something new had to be devised, and over the winter of 1857–58, Thomson devised it.

He called it a mirror galvanometer. Galvanometers, devices for detecting electrical currents, had existed for years; Vail had used one. But Thomson's galvanometer consisted of a tiny mirror weighing only a grain of sand and attached to electromagnets even smaller that would react with the electric telegraph signal. A candle was placed a short distance from the mirror, providing it with light to reflect. Between the mirror and the candle was a calibrated scale resembling a child's ruler. As the message came in — slowly, at six words a minute — the electromagnets moved the mirror, the mirror reflected the candlelight onto a point on the scale, and one of the operators sang out the letter of the alphabet thus indicated. With Whitehouse's receiving apparatus, modeled on Morse's telegraph, it was a wonder the message had arrived at all. And the mirror galvanometer could do more. When the little spot of light flew completely off the scale, it would reveal a fault in the line — not necessarily a break, but a place where weak insulation would soon cause a break. A mathematical computation would then tell the engineers, approximately, where the fault was, and it could be corrected (although not easily). Not without a struggle, the mirror galvanometer would become the standard oceanic reception device for the most of the rest of the century. Using it, the speed of transmission over an undersea cable was 3,700 miles a second. The battery that powered the entire cable was the size of a thimble. Daniell batteries, such as Morse had used, were about the size of a milk bottle.

Other changes were afoot. After *Niagara* and *Agamemnon* rendezvoused at Plymouth on March 9, 1858, there was a training period in the Bay of Biscay using salvaged cable from the year before. Then the two vessels proceeded to midocean. This year, the advice of the engineers was followed, and the ships would make a splice, *Niagara* would proceed to Newfoundland, and *Agamemnon* would make for Valentia. But far before midocean, the older salts began to examine the birds wheeling above the wake. Some of them were stormy petrels, believed to embody the souls of drowned seamen come to warn their comrades of dirty weather ahead.

It was the worst storm anyone aboard had ever experienced, and it went on for days. The escort vessels scattered and disappeared. *Niagara*, with her superior sea-keeping qualities, edged in close to *Agamemnon*, standing by and helpless. *Agamemnon*, with a London *Times* reporter embarked and recording her ordeal, repeatedly rolled through a hundred-and-eighty-degree arc, now down on her beam ends to starboard, now to larboard. Captain Preedy faced her into the wind and ran before it. Nothing helped. The crew could not keep the deck. Below, a marine grasped what he believed to be a wooden ledge, only to find his fingers agonizingly crushed when the seam closed again. A main beam broke and was held together with screw jacks. The on-deck coal broke loose; avalanches of coal poured down the hatchways. The huge coil of on-deck cable began to unwind. Only superb seamanship — or a miracle — could save her now.

When it was over at last, *Agamemnon*'s sick bay was full to overflowing, one man was dead, and another had lost his mind. The telegraph fleet sailed on. The next day, at midocean on June 26, the splice was made; the two cables were threaded into an eight-foot-long, three-hundred-pound iron-sheathed wooden cradle shaped like a shallow boomerang, a good-luck sixpence was inserted into the wires, a weight was attached, and the spliced cable was sent to the bottom of the sea. The ships proceeded toward their separate destinations. At three miles apart, *Niagara's* cable fouled again in its improved paying-out machine and broke. The ships rejoined. Another splice was made, and they began again. At three in the morning of June 27, with the two ships eighty miles apart, the cable went dead. Aboard *Agamemnon*, a blue light was burned and a cannon was fired. The ships rejoined again. No one knew what had happened. Perhaps the cable was trying to tell them something. Ashore and at sea over the course of the last year, most of it had been loaded, coiled, offloaded, recoiled, loaded again, and coiled for a third time. *Agamemnon's* deck cable had gotten into an ungodly tangle that took hours to set right. Yet, whenever they tested the cable, it seemed sound. But no one had ever tried to build a railroad to the moon. They still didn't know what the rules were.

The cable was spliced again, and the ships set out. *Agamemnon* built up speed to two knots, then four. On *Niagara*, Captain Hudson forbade conversation. A hundred miles slipped by, then two. Then the cable went dead. It had broken twenty feet astern of *Agamemnon*, now four hundred miles distant, under brake pressure of a thousand pounds, well within the limit of normal tolerance. The cable should not have broken.

Agamemnon returned briefly to midocean, *Niagara* returned to Ireland, and while the two ships were sorting themselves out, Field went to London and met with the board of directors. Three years ago he had told Lord Clarendon, the British foreign secretary, that if the cable broke, he would get another one and start again, and he proposed to do so now. The chairman of the board suggested winding up the enterprise and selling the cable to recover a little money. The vice chairman characterized the whole undertaking as sheer madness, left the room, and resigned. Field pointed out that there was still enough cable in the holds of the ships to complete the job in the current season. Somehow he carried the day.

Part Four: The Third Transatlantic Attempt

This time, there were no crowds and no cheering. As the ships left Ireland for the third time, the London *Times* correspondent recorded, "the squadron seemed to have slunk away on some disreputable mission than to have sailed for the accomplishment of a grand national scheme." The attempt, another correspondent wrote, was "a mad freak of stubborn ignorance."

This time, the job was a piece of cake. Midocean again, splice, and sail. At sunset the cable fell silent for an hour and a half, although Thomson's galvanometer showed no break in the insulation. Then it came alive again. The next day, the ship's officers found that *Niagara* was sixteen miles out of position. The weight of metal she carried had deflected the compass, and one of the escorts was sent ahead to lead the way. Otherwise, day after day, night after night, the paying-out machinery thrummed steadily, and there were no further incidents.

On *Niagara*, that is. *Agamemnon*, an unlucky ship, again had the more adventurous time of it. A fault was found in the cable in the hold. With only twenty inadequate minutes to make repairs, the ship was stopped, the cable was carefully held by the brake while the crew held its collective breath, and the splice was made. The ship was running against the wind and burning coal at a dangerous rate. Captain Preedy had visions of burning his spars and masts, then his furniture and his decks. A whale came close to the cable and sounded out of sight. In the middle of the ocean's empty vastness, two passing Yankee ships, on separate occasions, took it into their heads to approach *Agamemnon* on a collision course. Only a vigorous banging of cannon from the escort seemed to deflect them from their path. The weather turned nasty again. But the cable continued to play out; they had the rhythm now. At first light on August 2, they raised the headlands of Ireland and woke the population of Valentia with their guns. The local nobleman, the Knight of Kerry himself, met them as they came ashore.

On the other side of the ocean, at one forty-five in the morning of August 5, 1858, *Niagara* slipped anchor in Bay Bulls Arm, Newfoundland. Field immediately went ashore to alert the men in the temporary telegraph house that had been erected there, and at six, half a dozen ship's boats, paying the cable out behind them, landed the cable on the beach. First Lieutenant North handed it to Captain Hudson and, with the officers from the British escorts leading and the cable crew of *Niagara* behind, they made a procession. Within half an hour, the cable was connected with the line to New York.

In America, pandemonium ensued for days on end. A hundred guns were discharged on Boston Common. San Francisco paraded its firemen, accompanied by horse-drawn floats. Henry Wadsworth Longfellow took his children out to see the festivities. As an achievement, the cable was likened to the discovery of the New World and the invention of printing. Field was compared to Christopher Columbus and Johannes Gutenberg, and a large volume of appalling poetry was composed in his honor.

> Come listen to my song;
> It is no silly fable;
> 'Tis about the mighty cord
> They call the Atlantic Cable.
> Twice did his bravest efforts fail,
> And yet his mind was stable,
> He wasn't the man to break his heart
> Because he broke his cable.
> And may we honor ever more
> The manly, bold, and stable,
> And tell our sons, to make them brave,
> How Cyrus laid the cable.[4]

In New York, documents pertaining to the cable were entombed in the cornerstone of the new St. Patrick's Cathedral, and the workmen building Central Park, evergreen sprigs in their hatbands, filled Broadway from Union Square to Fifty-ninth Street in an unauthorized procession. Such were their numbers, wrote *New York Tribune* editor Horace Greeley in what one suspects was a rather dry mood, that "if all the men and teams in this turnout are kept at the city's work we shall soon have a great improvement in the new park." The city's firemen turned out at night with their specialty, a torchlight parade. Cannon were fired, City Hall was illuminated, and its celebrated cupola caught fire and burned down; the building itself was narrowly saved. Field, unlike Morse, went out of his way to thank by name the British and American officers and the British technicians who had made the cable possible.

The British were amused. They had detected, they said, a certain amount of Britishlike "drive" among the Americans after the cable broke for the second time, and much was made of the fact that the cable had reversed the verdict of the Revolution by making England and America one country again. Bright and Thomson were knighted, lesser honors were dispensed, and regret was expressed that Field, who had become immensely popular, was ineligible for one by reason of his nationality. For years thereafter, as he traveled back and forth across the Atlantic, he was affectionately called Lord Cable and regarded as the calm, true voice of American public opinion.

In England and America, Queen Victoria and President James Buchanan sat down to compose their inaugural remarks. But there was a delay. The technicians were still testing the line, Field explained. The technicians themselves were puzzled. Sometimes transmission was so clear that Newfoundland and Valentia could exchange pleasantries about the weather, and sometimes it was so garbled that the message log consisted of repeated requests for a repetition of the last message. When the chiefs of state were finally allowed to exchange pleasantries, the queen's message of 98 words took sixteen hours to transmit; the president's 149-word reply took ten hours. Professor Thomson was alarmed. He should have been. Something was seriously wrong. Whitehouse, still the cable's chief electrician, was trying to prove a point: that his Morselike transmitting apparatus and huge array of batteries were right and Professor Thomson's mirror galvanometer was all wrong. To prove this point, he was pouring gigantic amounts of electricity down the cable.

It is easy to see why he thought so. Like Morse, Whitehouse was a man of sturdy common sense. He was a man of affairs, he declared, and he knew how the job was to be done. He also knew how it was *not* to be done: with trickle power and grain-of-sand receivers activated by a battery no larger than a lady's thimble. It was nearly 1,700 miles from Newfoundland to Ireland, closer to two thousand if you counted the actual amount of cable that had been laid. Whitehouse deployed huge five-foot induction coils that generated 2,000 volts of electricity from a vast array of batteries, and used modified landline receivers of proven quality, though proven only when the lines were on land. Moreover, Whitehouse seemed to be vindicated. As the days passed, the clarity of the signal suddenly and dramatically improved. Secretly, to save face while pretending that his equipment actually worked, Whitehouse had begun to use Thompson's mirror galvanometer. But it was too late to save the cable.

September 1 was a typical day for Field. At noon he and the officers of the telegraph fleet landed at Castle Garden at the foot of Manhattan and received a 21-gun salute. Accompanied by the British minister, military and naval officers, and a hundred veterans of the War of 1812, he then participated in yet another monster parade that ended at the Crystal Palace on Forty-second Street and Fifth Avenue. Speeches followed, gold medals were awarded, and at some point in the proceedings, Field received a telegram. "C.W. Field, New York," he read. "Please inform . . . government we are now in position to forward — ." And there the message ended. It was the last coherent telegram sent over the cable.

Thomson replaced Whitehouse as chief electrician at Valentia, but nothing could be done. The cable wasn't dead, exactly, but it flickered, it faded, and it was too weak to be of any use. And, like the moon landing of 1969, voices were now raised to say that it had never happened. The astronomer royal reminded everyone that he had known all along it wouldn't work. More than one reporter wondered when Field had sold his shares in the cable and how many he had sold. President Buchanan mused that he had transcribed his remarks well in advance, and he supposed the queen had done so too. It was in vain, at least for a while, that Field's supporters pointed out that Cunard had reported a ship collision over the cable in real time, something that was impossible to do by copying the news later. They also tried to prove that the cable had worked by remarking that the death of a popular telegrapher had been learned of on both sides of the Atlantic *when it happened*, and that breaking news in London had been received in New York the same day. The American skeptics were particularly intrigued by the fact that the Newfoundland end of the cable had been manned by Charles de Sauty. It didn't sound like a real name, they said.

Estimates of the number of messages actually sent over the cable differed by a degree of magnitude. Some counts were as high as four hundred, which worked out to a hundred a week, but the number 271, compiled from the operating logs by George Saward, the Atlantic Cable Company's secretary, was probably correct. The handling of the cable — the onloading, the offloading, the storage, the problems on *Agamemnon* before and immediately after the storm — were reviewed. A close look was taken at the actual operation of the cable. It was concluded that when Whitehouse used his own equipment, only four intelligible words had been received, the longest of them "be." It was also concluded that Whitehouse's inappropriate equipment and huge currents had burned through the cable's insulation. "He sent a stroke of lightning over the cable," says one of Field's biographers, "which only required a spark."

But for the methodical British, fixing the blame was not enough. They were determined to lay a great many more cables, including a new Atlantic cable, and a court of inquiry set out to find the right way to do it.

PART FIVE: THE CIVIL WAR INTERLUDE

Field, meanwhile, was distracted by the part of his life that paid his bills. A financial panic in 1860 threw his wholesaling company into bankruptcy, his warehouse burned down, and to stay afloat he was required to mortgage everything he owned. He was not a ruined man — he settled with his creditors for twenty-five cents on the dollar — and he was not poor, but more than ever his fate was bound up in the cable project. He had sunk too much money into it, and the only way to recover his investment was to lay another one.

Still, events intervened. Field was one of two escorts who brought candidate Lincoln to address an important audience at Cooper Union. Later, after the South seceded, he was absorbed in putting certain land telegraphs (in which he had invested) onto a war footing, and even when he returned to England he found himself — as the voice of American opinion — deeply involved in the *Trent* affair, after a Union gunboat, in defiance of international law, removed two England-bound Southern diplomats from a British ship in international waters. The diplomats in question were the same Mason and Slidell who had given him so much trouble in Congress.

The popular Field, always able to have his day in the London press, pointed out that the cost of Britain's warlike preparations had cost more than an Atlantic cable, which would have done much to settle the crisis. But the lack of a cable was a mercy, in a way. Although Field had to wait fourteen agonizing days to discover the outcome of Gettysburg, he was spared for weeks the news that Federal troops, arriving in the nick of time, had cut down a large party of draft rioters only feet from his house on Gramercy Park, where his wife and children were. Field was well known to the rioters as a staunch Union man.

And, despite the British government's pro-Southern proclivities, he was well known and well regarded in Downing Street and Whitehall. The court of inquiry had been seated to look into the cable's failure (it blamed Whitehouse) and its recommendations were in. It had found unanimously in favor of a new Atlantic cable. The British government still looked on the project with favor, although it would obviously be postponed pending the resolution of hostilities in America. The recent failure of a heavily subsidized Red Sea cable to India still smarted in England — the hard-pressed British taxpayer would pay for it for years to come — but in England and Europe, if not in hard-pressed Washington, undersea telegraph cables were still the wave of the future. Glass, Elliott, the former pit rope company, had turned itself into a cable-laying company and had laid a line from Malta to Alexandria; India was reached despite the Red Sea fiasco, and the French had a line to Algeria through waters more than two miles deep. In view of these triumphs, an Atlantic cable still seemed feasible and even desirable as England pressed onward in its quest to be the center of a world communications network. For its part, the British government was still willing to subsidize such a cable with up to $100,000 a year, and it would guarantee an 8 percent return on $300,000 of paid-in private capital for twenty-five years — *provided the cable worked*. Field had no doubt that the cable, properly fitted with tiny batteries and Thomson's mirror galvanometer, would work. He gathered new estimates — the British court of inquiry had laid down a series of specifications based on its experiments and recent cable-laying experience — and returned to America for a fresh infusion of capital.

At home, again, he ran into a stone wall. Everywhere he went — Boston, Providence, Albany, Buffalo, Philadelphia — it was the same. Field was cordially received and politely attended, and got nowhere. To the conservative American investors, many of them bloated with money from war profiteering, a British-subsidized Atlantic cable with a guaranteed return was closely akin to the South Sea Bubble, the early-eighteenth-century English speculative frenzy that had seriously lightened the purses of the royal family and the aristocracy. Meanwhile, time seemed to be running short. British investors were actively exploring a transatlantic route through Iceland and Greenland, short and easy hops, and the

American Perry Collins was building an Alaska-Russia route with an army of Siberian laborers. With only $350,000 in new capital, most of it raised in New York and most of it the monetary equivalent of a sympathy vote for Cyrus Field, Field returned to England. It was his only hope.

The news in England was also not good. John Brett, who had done so much to raise the money for the first attempts, had died, and the old enthusiasm for Field's project had flickered out. As in one of those Victorian melodramas that so much of the Electric Revolution resembles, things looked dark indeed both for the cable and for the personal fortunes of Cyrus Field. But it was the darkness before the dawn. Field encountered Thomas Brassey, who had gotten in on the ground floor of the railway boom, had built them in England, France, and Canada, and had become the most prominent railroad contractor in the world. Brassey told Field to put him down for a tenth of whatever the cable cost. The people around Brassey said he had a way of doing things like that, and sometimes his investments even paid off.

With Brassey aboard, it was as though Field had hurled himself with all his might through a paper door. Glass, Elliott had merged with the Gutta Percha Company, forming an integrated enterprise that manufactured cables and laid them. Now the new company, which eventually became Cable & Wireless, came forward with a stunning offer. It would buy all the remaining stock in Field's new venture, and it would buy a further $500,000 of the venture's bonds. In addition, it expected to bid on the job, it did, and — now styling itself the Telegraph Maintenance and Construction Company — it won the contest in a walk. It stood to gather in most of the British government's guaranteed return, and would get paid for its work no matter what happened. Dynamic England was not cautious America. It knew how to get things done.

PART SIX: THE FOURTH TRANSATLANTIC ATTEMPT

The new cable showed the benefits of experience. It still had seven
redundant strands of copper wire, but following Bright's and Thomson's recommendations it was thicker by more than half an inch.
The core was first insulated with a new miracle substance, Chatter-
ton's Compound, a mixture of gutta percha and Stockholm tar, and
then with four layers of undiluted gutta percha. Like the cables of
1857 and 1858, it was next armored with wire. Then — another
new development — it was covered with hemp. The hemp would
protect the armor from rust, and it would add buoyancy. The old
cable could support itself in five vertical miles of water, but the new
cable could support itself in eleven miles. Going into full fail-safe
mode, Field and his people, again including Professor and now Sir
William Thomson, decided to avoid the possible peril of splices by
laying the new cable in a single length. The question was how.

The answer came in yet another improbable bolt from the blue.
Years before, the great British engineer Isambard Kingdom Brunel
had suggested to Field that his newest creation, Great Eastern, was
the ideal ship to lay the cable, and now Great Eastern was available.

Great Eastern was by far the largest steamship built to date; she
would not be exceeded in length (693 feet) until Oceanic II was
launched in 1899, or in displacement (22,500 tons) until Lusitania
was built in 1906. Her two ninety-ton paddle wheels were fifty-
eight feet in diameter — the crankshaft was the largest piece of
forged iron in the world — and the propeller checked in at twenty-
four feet. Her engines developed a staggering 11,000 horsepower,
and it took two hundred men to raise her enormous anchor. She had
five funnels and six masts, each named for a day of the week, and
aloft she carried one of Humphry Davy's arc lights which "bathed
the ship in perpetual moonlight." She was also, even to a number of
Englishmen, a glaring example of British hubris at the height of its
dark and slightly preposterous powers. Great Eastern was designed to
carry four thousand passengers to Australia without stopping to take

on fresh coal, although she actually spent her short, commercially disastrous life as a passenger ship in the Atlantic trade.

Despite these impressive credentials, *Great Eastern* was nothing but bad luck. On her first voyage, she carried a crew of 418 and only 35 paying passengers. She was overengineered, her engines burned too much fuel, and she never made a dime for her owners, of which, by 1864, she had had three. Field, always eager for new experiences, had been aboard her during one of her trips to America. The food was execrable, the crew was drunk, there was nothing to drink but champagne. At the first opportunity after she reached America, he had gotten off and returned home by train.

Now *Great Eastern* was tied up in the Thames, looking for work. She was the only ship in the world that could carry the entire Atlantic cable, with storage space for hundreds of miles more in case of accidents. And the head of her latest group of owners, a man with the Dickensian name of Daniel Gooch, had come forward with an offer. There was something about the tireless and determined Mr. Field that appealed to him, something sporting, something rather English, in fact, and Daniel Gooch was something of a sporting man himself. Like Thomas Brassey, he had made his money in the railroads. He had been a locomotive superintendent at the age of twenty-one, had given Queen Victoria her first train ride, and was now the chairman of the Great Western Railroad as well as the head of the consortium that currently owned *Great Eastern*. And he had a proposition. If his ship laid the cable, he expected a payment of $250,000 in cable stock. If *Great Eastern* failed to lay the cable, there would be no fee. Field accepted. Cyrus Field, wrote William Russell of the London *Times*, the reporter invited aboard to record the voyage, "seemed bent upon solving the problem of [*Great Eastern's*] existence." Actually, it's hard to see what choice he had. No other ship could carry the cable.

This time, the Americans did not contribute a naval presence. During the hostilities in America, the British had imposed a "twenty-four-hour rule" that effectively barred Union warships from

British ports, and denying an American naval escort to the cable fleet was retaliation. If the British wanted an Atlantic cable, the American government in effect said, let them build it. And so they did. The Atlantic cable was now almost entirely a British venture.

As the cable was manufactured, it was stored in specially prepared water tanks. It was transported to *Great Eastern* in old naval hulks — cut-down, superannuated ships of the line — and winched aboard covered from the damaging rays of the sun by tarpaulins. It was then coiled in the vessel's three holds and stored under two thousand tons of water. The Prince of Wales visited, a sign of royal approval. Eight thousand tons of coal were laid on. Field bought a cow and a dozen oxen, twenty pigs, a hundred and twenty sheep, flocks of geese, ducks, and turkeys, five hundred chickens, and eighteen thousand eggs. It was said that *Great Eastern* resembled a gigantic floating farm. A newly designed paying-out machine was installed, and in the bow there was a steam engine and a crane to retrieve a broken cable. The main deck was called Oxford Street, because of a double row of lights that had been installed to illuminate the work at night. De Sauty was placed in charge of the telegraph room, with Professor Thomson again standing by to offer advice. Field dressed for the voyage in a deerstalker cap and an Inverness cape. On July 15, 1865, they set sail on the fourth attempt.

With slight modifications, the drill was the same as before. First to Ireland to land the inshore cable, a much heavier version of the oceanic one, designed to guard against anchors, nets, and damaging currents. There was a splice at sea, and the new cable fleet — *Great Eastern*, two British warships, and two ships hired by Field — turned westward on July 23, a Sunday. Sailors were superstitious about Sunday sailings, but if the schedule was kept, they would arrive in Newfoundland under a useful full moon. Cyrus Field was the only American embarked.

None of them, not even Captain James Anderson, borrowed for the occasion from Cunard, had ever been aboard a ship so stable. Accustomed to the sharply rolling craft of the era, they swore she didn't move with the sea at all; for the first time in all his voyages,

Field wasn't seasick. At midnight, seventy-three miles out, a gun sounded. A fault had been detected in the line, approximately ten miles astern. There was no choice but to haul the cable up and cut the damaged section out. The cable was severed, worked forward over an eighth of a mile of cluttered deck, and the first unforeseen problem was discovered: the winch in the bow was underpowered for the job. It took them until Tuesday to bring up the fault out of shallow water. Somehow, it seemed, a length of iron wire had been inserted in the heart of the cable. It was about as long as a nail. No one knew what to make of it.

Still, they had their spot of fun. They could thank Billy Russell of the *Times* for that. He was the first modern war correspondent, more or less making up his profession as he went along. He covered the Crimea, where a nervous high command banished him to Bulgaria and he defeated the military mind by using the new Black Sea cable to write his dispatches, and he covered the American Civil War, where Lincoln gave him the name "Bull Run Russell" because his painfully accurate reporting of the Union catastrophe got him thrown out of the country. Now, on *Great Eastern*, he published the first shipborne newspaper ever written, the *Atlantic Telegraph*. The cable kept him in touch with Ireland, and Ireland kept him in touch with Europe and North Africa thanks to the undersea lines that had been laid during the last decade and a half. Russell compiled a daily bulletin of foreign news and stock reports, reported planned shipboard activities, and composed volumes of verse that he, unlike the usual run of telegraph poet, knew was as awful as it sounded.

> Under the sea! Under the sea! Here's what de Sauty is saying to
>   me,
> Such testing as this is the perfectest bliss! Insulation is holding
>   up strong,
> So we'll test! test! test! with coils and rheometers, keys and
>   galvanometers!
> Test! test! test! Test each minute all night and day long.
>
> (*chorus*)

Copper and zinc! acid and stink! tink-a-tank-tink-a-tank-tink-
a-tank-tink.
Copper and zinc! acid and stink! success to continuity.[5]

While Russell was having his fun, an old finding was confirmed.
The transmissions over the new cable, like the old, were improved
by the depth and cold of the sea.

On July 29 at 1:10 in the afternoon, the cannon was fired again.
There was another fault in the line. It took them until ten that
night to bring it up. Another piece of wire had been driven through
the cable.

They drew the obvious conclusion: the cable had been sabo-
taged. Bright metal was visible where the offending wire had been
snipped, and it was noted that the same cable watch had been on
duty when both faults occurred. Under a cloud, and with the rest of
the crew muttering darkly, the presumed offenders were assigned
other tasks, and the gentlemen of the ship — Field, the technicians, and the officers — were assigned surveillance duties.

If it was sabotage, it wouldn't be the first time. In a well-known
case, a crewman on a North Sea cable-laying expedition had been
bribed by the competition to drive a nail through the line. Later,
Field and the other ship's gentlemen decided, on what seemed like
good evidence, that no human agency had been involved at all,
that a piece of the cable armor had somehow worked its way in and
broken off and no skullduggery was afoot. It was, nevertheless, a
persistent truth of the Electric Revolution that workmen deliberately ruined equipment. In later years, the American arc light magnate Charles Brush found himself traveling miles to remove a tack
or a cotter pin that had been tossed into the coils of one of his dynamos, and as late as the mid-twentieth century, telephone workers
made a joke of throwing their screwdrivers into the equipment.
There was something about electricity — its invisibility, perhaps —
that inspired costly mischief.

Field, decked out in his usual Holmesian array of deerstalker
and Inverness, began to haunt the cable tanks. Although Captain

Anderson had disconnected the paddle wheels and was using only the propeller to keep from outrunning the cable machinery, they were making a good speed, a hundred and twenty miles a day. Already they had passed the midocean spot where the cables of 1857 and 1858 had been lost. The contact with Ireland remained clear. Russell continued to publish his newspaper. The new machinery worked faultlessly. They were only eight hundred miles from Newfoundland in the last days of July when Field, standing guard in the cable tank as usual, heard a grating sound and saw a flash of metal in the uncoiling cable above his head. Another wire had worked loose. Too late, he tried to raise the alarm. The infected length had already gone over the stern.

This time, the cable was two and a half miles down. Once more, the cable was cut and hauled to the bow, a task that took nearly ten hours. The underpowered retrieval engine was fired up. Captain Anderson cut the main engines. There was no way a ship the size of *Great Eastern* could dance lightly over the waves, backing and advancing as the cable was pulled aboard. And so she began to drift. Drifting, she began to chafe the cable against her hull. Nothing could be done. *Great Eastern* simply could not maneuver; she was as long as a city block.

The officers and gentlemen were at lunch when a technician, actually staggering under the burden of his news, reported that the cable was gone; once again, it seems that there was more truth to the melodrama than we now suspect. But they were old hands at this by now. Field retained his usual and infectious optimism in the face of the usual catastrophe; Russell claimed that whenever a cable broke, he immediately sat down to write a prospectus for the next one, although in private Field allowed himself a thought that was perhaps revealing. "It was this pin's point which pricked the vital cord," he wrote later, "opening a passage through which the electricity, like a jet of blood from a severed artery, went streaming into the sea." The cable, with its repeated failures, was in danger of bleeding him white.

In place of the knighted Charles Bright, who was otherwise

engaged as a member of Parliament, the chief engineer of the current venture was Samuel Canning, a veteran of *Agamemnon*. He had already laid his plans for just such a contingency. He proposed to grapple for the lost cable, bring it to the surface, and continue.

Unfortunately, the grappling equipment was on HMS *Sphinx*, one of the escorts. *Sphinx* had gotten separated from the telegraph fleet during a spot of dirty weather, and had not returned. There were, however, five miles of wire rope aboard *Great Eastern*, and a set of grapnels resembling five-pointed anchors, which were part of the ship's inventory. The wire rope, intended to lash the cable to a buoy if the eventuality arose, was in a number of pieces. Canning set the crew to work piecing together two and a half miles of it with metal connectors called swivels, and Captain Anderson set to work positioning *Great Eastern* for the attempt.

They knew with some precision where the broken cable was; Anderson had been fixing the ship's position at just the moment when it was lost. The problem, again, was *Great Eastern*'s lack of maneuverability. When the rope was spliced and the grapnel attached, it was thrown over the side for its two-hour descent. Anderson positioned the ship three miles south of the cable line. Then he allowed the ship to drift down with the wind and current. They were trying to pick up a strand of spaghetti with a bent pin at the bottom of a thousand-foot well.

The work went on all day and night on August fourth, fifth, sixth, seventh, eighth, and ninth. They got the cable. They got it twice, in fact. The first time, they raised it three-quarters of a mile before a swivel broke. The second time, they raise it by a mile and a swivel broke again. They marked the spot with a buoy and took stock of the situation.

There was still a quantity of wire rope, but not enough to reach the bottom. There was also a large quantity of hempen rope and Manila hawser, but they were divided into twenty-five lengths of a hundred fathoms each. And the underpowered steam winch could no longer be relied on. On the night of the ninth, as part of the crew sat down to piece together two and a half miles of improvised line, a forge was set up on *Great Eastern*'s forward deck. Using iron plates, they were going to increase the diameter of the capstan by four feet. If the steam winch failed, they would bring up the cable by hand. They worked through the night, with the fire of the forge glowing and flaring eerily against the fog.

The first pass failed. Captain Anderson returned to the south and drifted north. The cable was grappled and pulled up to eight hundred feet off the bottom. Then the pieced-together rope broke. There was nothing for it but to go home. On their arrival, to their considerable astonishment, they were greeted as heroes.

By now, no one in England doubted that the thing could be done. Transmissions over the cable had been flawless to the last, the machinery (except the steam winch) had worked as advertised, *Great Eastern* was the ship for the job, and they had all been brave men and true. Daniel Gooch kicked in £20,000 and signed on for the next cruise. Thomas Brassey pledged another 10 percent. Junius Morgan, the American financier who lived and worked in London, lent a hand. After a brief delay caused by a tussle with British financial law, the money — a fresh $3 million — was raised.

Field returned to New York and almost immediately sailed back to England when a message from Captain Anderson arrived, saying that nothing seemed to get done unless Field was on the spot. He traveled back and forth so often, it was said, that he seemed to be transmitting himself by electricity, like a message on his dreamed-of cable. By now, he had spent a quarter of his life trying to bring it to reality. There was the matter of fair play involved, the British thought. After all this time, he deserved to succeed. He had earned it.

PART SEVEN: THE FIFTH TRANSATLANTIC ATTEMPT

The new cable — the fourth: the cable on the third attempt was a leftover from the second one — was much like the previous one, but the armor sheath was galvanized, both to prevent rust and to guard against another break and puncture. Just to be on the safe side, Captain Anderson ordered pocketless coveralls for the crew, and they were ordered to wear them at all times. A pocket could conceal a cutting tool, and Anderson was not a man to take chances. The steam winch was refitted with two 70-horsepower steam engines. *Great Eastern* sailed on June 30. Once again, Cyrus Field was the only American aboard. The splice with the inshore cable was made on Friday, July 13, 1866, a date of ill omen, and for the fifth time a telegraph fleet headed west. On July 15, on a night of rain, the cable fouled in the after-tank. It took two hours to untangle.

At night, a red lamp marked each mile as the cable came up out of the hold. By day, a flag was used. Captain Anderson proceeded at a steady five knots. Midocean was passed. The Grand Banks were reached under a cover of fog. HMS *Terrible* sounded two blasts on her horn; *Medway*, a hired ship, sounded three; *Albany*, another hired ship, four; *Great Eastern*, one. After all this time and heartbreak, it would be the ultimate fiasco to sever the cable in a fog bank. Land birds and floating branches appeared. On the afternoon of July 27, *Terrible* signaled that HMS *Niger*, the guide ship that was to meet them by prearrangement, was in sight. The following morning, *Great Eastern* dressed her yards, fired a salute, and dropped anchor off the Newfoundland fishing hamlet of Heart's Content. Before the cable could be landed and connected with the specially built telegraph house, London sent the headline from that morning's *Times*. "It is a great work, a glory to our age and nation, and the men who have achieved it deserve to be honored among the benefactors of their race. Treaty of peace signed between Prussia and Austria."

Messages of congratulation poured in. One came from San Francisco, a city that could now reach Bombay by telegraph at a rate of eight words a minute and a price of a hundred dollars for a five-word message. Another message was from Ferdinand de Lesseps in Alexandria, Egypt, where he was putting the finishing touches on the Suez Canal. Later, demonstrating the technology, one of the electricians telegraphed England with a battery made from a percussion cap, a tiny sliver of zinc, and a drop of acid. Legend has it that the liquid matrix was, instead, a human tear.

The cable fleet returned to midocean and grappled for the cable of the year before. Thirty attempts were made. The captain of *Albany* collapsed from overwork and was confined to sick bay on *Great Eastern*. The captain of HMS *Terrible* put his men on half rations, ran low on coal, and returned to Newfoundland. When the cable was finally hauled aboard *Great Eastern*, utter silence prevailed. When a signal was sent and Ireland answered, rousing cheers were given. Cyrus Field went to his cabin and wept.

PART EIGHT: AFTERWARD

It was possible to make too much of all this, and Americans did. To the British, ocean telegraphy was by now old hat. There were many and distinct advantages to an Atlantic cable. For example, it had previously taken weeks and sometimes months for an American merchant, investor, or speculator to learn the price of a commodity on the London exchange, the center of world finance. Business could now be done in real time at a steadily decreasing cost. Cable transmission never became cheap, but the price was soon reduced from $100 for a brief sentence and the speed of transmission was gradually ramped up to fifteen words a minute. And, as Field and Morse had predicted, Britain and the United States never went to war again, although it may be saying too much to give the total credit to the Atlantic cable.

In matters of colonial administration, as in the affairs of business, the British were already familiar with the advantages of telegraphy. Ocean telegraphy added a whole new dimension to the matter, as the British had discovered in a slightly different context during the Crimean War, where they had also discovered, among other things, that the cable made a legend of Florence Nightingale, brought major reforms to military medicine, and vastly improved the lot of British troops, who had previously been treated little better than slaves. With the oceanic cable, Britain's imperial satraps lost much of their exhilarating freedom of action. Out of touch with the War Office, the Colonial Office, and Downing Street, governors general and military men in distant outposts had resembled the rulers of independent countries, reacting to fast-moving events, making independent decisions, and sometimes disobeying explicit orders. In the 1840s, General Sir Charles James Napier had been specifically instructed not to occupy the Indian province of Sind. He did it anyway, and sent the following dispatch to London: *Peccavi.* "I have sinned." There had been no way to stop him, or even to know what he was doing. The oceanic cable changed all that. Ambassadors, too, found that they had lost much of their discretion, and younger sons dispatched to manage the family's affairs in distant climes discovered that they were actually expected to work. Britain, with its monopoly of gutta percha and its hard-won expertise in cable laying and manufacture, became the center of world communications, a position it would not relinquish for a hundred years.

Daniel Gooch received a baronetcy, Captain Anderson was knighted, and *Great Eastern* went on to other adventures. She was bought by Napoleon III and pressed into service as a transatlantic liner to bring visitors to his Paris Exposition of 1867. Four men were killed raising her anchor. Later, a storm struck and the passageways were flooded. The social director of the ill-fated voyage was a young writer named Jules Verne. Later, the ship was pressed into service laying another cable, this one from France to America, a task she successfully completed. She was then turned into an amusement park, did not thrive in her newest role, and was scrapped in the 1880s. It took two years to break her up, and a human cadaver was found in her double hull. There had always been a legend about a Phantom Riveter, trapped in a sealed compartment, whose gun could be heard hammering away during the dark watches. The story, some of it, is true.

Cyrus Field toyed for years with the notion of laying a cable across the Pacific to Japan and China. But the only American ship ever to have laid a cable was *Niagara*, no one in the United States had any experience in manufacturing cable, and he gave up the idea when he saw the Pacific with his own eyes and realized how vast and deep it was. He bought a country house that made Jay Gould his next-door neighbor and turned his attentions from paper whole-saling to the running of elevated railroads in New York City, where he became known, because of his usual energy, as the "locomotive in trousers." Like Peter Cooper and other successful men of his generation, he believed that a man of wealth should also seek the betterment of his society while making a reasonable return on his money. The contemporary Brooklyn plutocrat Alfred Tredway White called the practice "philanthropy plus five percent," and overall it served the country well. Field's elevated railroads were ugly things, but they kept New York from strangling itself with congested roadways and too many people, and Field tried to make them as palatable as he could. He paneled his passenger cars with mahogany, beautified his stations, and fought his stockholders — including Gould and Russell Sage — to reduce the fare from a dime to a nickel. But as he grew older and the Gilded Age arrived, he convinced himself that a new economy had been created, and that the new economy was personified by Jay Gould, who was — not for nothing — called the Exterminating Angel of Wall Street. Gould was a man who defined the word predator. Nonetheless, as a front man for Gould, the amiable and public-spirited Cyrus Field became president of the Wabash Railroad. Then, imitating Gould, he decided to corner the stock of the elevated railroads he had so responsibly run. Unfortunately, he decided to buy the stock with borrowed money. Gould bided his time as Field went ever more deeply into debt, and then he cleaned him out.

Field's wife and one of his daughters died and another was confined to an insane asylum. His stockbroker son embezzled a fortune from his clients and also his father, and became an inmate of the same asylum. Field died, a ruined man, at his country home on July 12, 1892. In his last delirium, he imagined himself again on the bridge of *Great Eastern* — or abandoned ashore at Valentia as everyone else sailed away.

J. P. Morgan salvaged what he could of the estate for the survivors of the family.

The mirror galvanometer and shrewd investments in oceanic cable made Sir William Thomson a very rich man. He bought a country estate and he sailed the world in the *Lalla Rookh*, his 120-foot yacht. He computed the temperature of absolute zero, cowrote the physics text that was regarded as definitive by a generation of scientists, and became Lord Kelvin of Largs, acclaimed as the greatest British scientist since Faraday. Using heat theory to compute the age of the earth, he concluded that the earth was 90 million years old and that Darwin's theory of evolution was therefore wrong, because not enough time had elapsed. As president of the Royal Society, he declared that heavier-than-air flight was impossible. But in five cable voyages between 1857 and 1866, he was known as the "genius of electricity," and the mirror galvanometer was his triumph. And by reasoning counterintuitively and turning the electricity down, he had broken through the cable's power barrier. Well into the twentieth century, a marine telegraph cable would consume 1/25,000 of the electricity used by a common doorbell. His achievement, and the British takeover of Field's Atlantic cable, were the high-water mark of the British effort in the practical applications of the Electric Revolution. This high point would never come again. As some Britons had foreseen, their home island became the world's center of telegraphic communications, but the telegraphic cable was a dead end. The technology evolved, but not by much. The next development in the Electric Revolution was about to happen elsewhere, and the British would not be there. They would never pioneer a new electrical development again. Instead, the Americans were about to take center stage.

CHAPTER SIX

MERLIN

It should be understood that Mr. Eddison does not belong to himself. He is the property of the telegraph company which lodges him in New York at a superb hotel; keeps him on a luxurious footing, and pays him a formidable salary so as to be the one to know of and profit by his discoveries. The company has, in the dwelling of Eddison, men in its employ who do not quit him for a moment, at the table, on the street, in the laboratory. So that this wretched man, watched more closely than ever was any malefactor, cannot even give a moment's thought to his own private affairs without one of his guards asking him what he is thinking about.

— *"This Astounding Eddison,"* Figaro, *1877*

In 1866, as the British concluded their triumph in the Atlantic, America outside the war-devastated South possessed, as it had since the days of Franklin, the highest standard of living in the world. Britain, however, was the world's great power, with the most efficient agriculture, banking system, railroad infrastructure, and government. Its navy commanded the seas, as did its commerce. It was the home and starting place of the Industrial Revolution. Only a short period before, half of the industry on the entire planet called Great Britain home, and its lead was still commanding. All over the globe, British engineering and British engineers ruled the roost. Its scientific establishment was equal to any and superior to most.

America, by contrast, was a second-rate power that had just torn it-self apart in a fratricidal struggle. But all that was about to change.

With the clarifying benefit of hindsight, it is easy to see what was about to go wrong for Britain and what was about to go right for the United States. At the heart of it was the British workforce. The Americans were the most literate society in the world. As long ago as Franklin's day, male literacy was close to seventy percent, highest in New England, where Boston female literacy was also seventy per-cent, lowest in the South, with the Middle Atlantic states some-where in between. Then, Britain's male literacy rate was in the vicinity of sixty percent, but while America continued to expand the schooling given to its yeomen and city dwellers, the British did not. Indeed, the British expended very little useful effort on the working class that comprised the bulk of the population. In Amer-ica, an historical accident guaranteed that the workers, by world standards, were very well paid: even in 1866, America was a thinly settled, largely agricultural country with an independent-minded populace and a small working class. In America, if you wanted workers, you had to beat someone else's bid. Britain, on the other hand, was thick with people, its agricultural and cottage industry sectors were shrinking, and the workers — drawn from what was called the surplus population — were miserably paid. This created a dilemma. Britain's workers were not consumers in the modern sense. They paid for clothing, food, and housing, and they paid for very little else. Unlike the Americans, they were not members of what was then called the middling classes, and beyond a certain and very limited sense, the British populace did not consume much in the way of goods, services, and the products that were the fruits of invention. Home of the Industrial Revolution, it was also the home of invention. The spinning mills, the very heart of the Industrial Revolution, were constantly improved. Improvement meant that the mills produced ever greater quantities of cheap cloth, far more than the country's population could absorb. The solution to the problem was to export it. If the export market collapsed, Britain's new factories would collapse. But within this problem, there was an even deeper problem.

For the longest time in Britain, as in America, the world of invention was the world of the self-made amateur, where university-trained men like Morse were the exceptions that proved the rule. Britain's Richard Arkwright, who built the first of many spinning mills, was a wig maker. Faraday, of course, had started life as a bookbinder's apprentice, and Sturgeon of the electromagnet as a bootmaker. These latter two, with William Thomson and Joseph Henry, were laying the foundations of electrical science and, in a variation on the entrepreneurial suicide that began to bedevil the manufacturing sector, were also sowing the seeds of the destruction of the British electrical revolution. In America, Henry was the last of his kind for the longest time, a cutting-edge scientist whose seminal discoveries could be the basis of an invention; in America, the universities did not become the seat of electrical discoveries. But in Britain, Faraday and Sturgeon and Thomson, whatever they were at the beginning, were the founders of an academic dynasty. British electrical science, in other words, was walling itself off from the British populace, and because the British were making no effort to create a modern consumer society, there appeared to be no market for the products inspired by British discoveries. America, by contrast, was still the land of the amateur, and it was also a society of consumers with surplus disposable incomes. When the two combined, the result was Thomas Edison, and the British blew one of the largest leads in scientific history.

Humphry Davy, the vivid and solitary Romantic scientist, had etched the outline of the lone inventor, Thomas Davenport had filled it in, and Edison embodied it through all his long life. As flexible and manipulative with the truth as Morse, he uncannily sensed the latent expectations of the public. For starters, he came from the right sort of family. And he equipped himself with the right sort of education, which was, he claimed, no education at all.

Once, every schoolboy knew the story of Edison, his deafness, his rolling laboratory, his trainborne newspaper. Tactfully omitted from the account was the fact that his father, a descendent of American Loyalists who fled to Canada after the Revolution, had returned to America on the lam from the British after joining the ill-starred Mackenzie Rebellion. The Edisons, it was said, had a way of joining the wrong side. Edison's father became a sometime storekeeper, sometime shingle maker, and sometime groundskeeper. His last and seventh child, Thomas Alva, was born in 1847.

Sometimes Edison said he had never gone to school a day in his life; at other times he laid claim to three months. The correct figure is four years, a highly respectable total for the time when America's high level of literacy was achieved with remarkably little schooling. During his formative years in Milan, Ohio, and Port Huron, Michigan, an American child was lucky if his education lasted twenty-four months, and almost no one but the scions of the wealthy attended secondary school at all. Given the nineteenth century's obsession with turning great men in their youth into fictitious monsters of virtue and humility, Edison's fudging of his schooldays seems a minor thing, a retrospective attempt to become one of the boys, but this does not account for the fact that his admirers also novelized his past in an attempt to fit him into a superheroic mold. For example, it was apparently an admirer who spread the story that his precocious reading included Gibbon's *Decline and Fall* when, in fact, one of his favorite books seems to have been Sylvanus Cobb Jr.'s adventure novel *The Gun-Maker of Moscow*. To Edison himself we owe the tale that he read Newton's *Principia Mathematica*, although with a typical Edison twist: that the book "gave me a taste for mathematics from which I have never recovered." Whether or not Edison actually read it is, at this point, probably as unknowable as it is unlikely.

Perhaps the truest thing ever said about the young Al Edison (he wasn't called Tom until the middle of his adolescence; the family pronounced their last name with a long *a* rather than a short *e*) was uttered by his father, Sam, who said his son had no boyhood days as others did; the lad was principally interested in machinery.

To Edison's contemporaries, this explained nothing. They could comprehend the machinery of the Industrial Revolution. They could see boiling water produce steam that drove a piston that, in turn, propelled a locomotive or powered the machinery in a factory. There was nothing very difficult about this, because every device, and every part of a device, had its analogy in the familiar and comfortable world. But the adult Edison dealt in electricity, and this was a very different kettle of fish.

The early Industrial Revolution was the last American time when an ordinary citizen was thought capable of understanding the natural world. Davenport had been forgotten, and not many people had known about him when he was alive. Morse was far from a simple yeoman, and Field was a New York businessman who had accomplished a feat with British money and British technology. The telegraph had been amazing, but for the bulk of the population it had been an amazing thing viewed from afar. It might have regulated the new railroads and begun to revolutionize commerce as practiced in the countinghouse, but these were fairly abstract phenomena. The bulk of the populace might live most of a lifetime without ever sending a telegram or thinking about why the trains ran on time. Not so Edison. His great years were the half decade between 1876 and 1882, when inventions ranging from the phonograph to the light bulb to the central power plant began to touch and transform the lives of everyone in the country. But Edison, like Conrad's Lord Jim, was emphatically one of us, a man of the people. He was not conveniently dead, like Franklin, so that his strange, incomprehensible achievements could be viewed with something approaching awe — and dismissed as having no relevance. He was not a lovable old man as Morse became. When he made his great inventions and achieved his great fame, Edison was in his thirties. He spoke like a man of the people. He told funny stories. And he did things with electricity, a force nobody, even Edison, understood, and nobody could figure out how the things he did with it worked. And so the young man of the people, Edison, had to become a child of superhuman genius. Like all superheroes who retain a shred of their humanity, he required an origin story, and vast but intuitive youthful brilliance was it. The alternative, he found, was repugnant. The alternative said he wasn't human.

It was not unusual that he began his independent working life at the age of sixteen, nor that he chose the telegraph as his profession. Given the life expectancy at the time — forty-five years — everyone started young. Port Huron was on the railroad, and where the railroad went, the telegraph went. There was a window of opportunity: with thousands of the country's telegraph operators in the service of either the Confederacy or the Union, jobs in the industry went begging, including a job in his own hometown. They were not, as it happened, the most desirable jobs in the world. And Edison, at least at first, was not a very good telegrapher.

* * *

Eighteen hours a day at the telegraph key had much the same effect as eighteen hours a day writing computer code. It was a solitary, self-referential profession with little human contact, and it did strange things to the mind. Telegraph operators tended to be slow to mature, fast to take to drink, and nomadic; few of them stayed in any one place for any great length of time. Wartime living conditions were primitive, the typical telegraph office was infested with vermin, and many telegraphers gave every appearance of hating their lives. Once, an Edison colleague named Billy L. arrived at the workplace at two A.M. full of hooch and attitude. "Suddenly the door was thrown open with great violence, dislodging it from one of the hinges," Edison recalled.

> There appeared in the doorway one of the best operators we had, who worked daytime, and who was of a very quiet disposition except when intoxicated. . . . His eyes were bloodshot and wild, and one sleeve had been torn away from his coat. Without noticing either of us he went up to the stove and kicked it over. The stove-pipe fell, dislocated at every joint. It was half full of exceedingly fine soot, which floated out and filled the room completely.

This produced a momentary respite to his labors. When the atmosphere had cleared sufficiently to see, he went around and pulled every table away from the wall, piling them on top of the stove in the middle of the room. Then he proceeded to pull the switchboard away from the wall. It was held tightly by screws. He succeeded, finally, and when it gave way he fell with the board, and striking on a table cut himself so that he soon became covered with blood. He then went to the battery-room and knocked all the batteries off on the floor. The nitric acid soon began to combine with the plaster in the room below, which was the public receiving-room for messengers and bookkeepers. The excess nitric acid poured through and ate up the account-books. After having finished everything to his satisfaction, he left.[1]

The next morning, surveying the wreckage, the manager announced: "If Billy L. ever does that again, I will discharge him."

Edison, by contrast, kept getting fired. He was unkempt, he preferred to spend his time reading or playing with the equipment, and his fellow telegraphers thought he was a hick, callow, immature, and — reading between the lines — rather too full of himself for everybody else's good. Among the telegraphers, his nickname was Looney.

In 1864 he moved his act across the border, to Stratford Junction, Ontario, where he invented a clockwork device that sent a "sixing signal" — an hourly tap of the key that told his superiors he was on duty — that enabled him to sleep during the night shift. He also put two trains on a collision course, the greatest sin a telegrapher could commit, and although he claimed extenuating circumstances, his supervisor was not impressed. So little was the supervisor impressed, in fact, that Edison hightailed it back to Port Huron, one step — or so he believed — ahead of the law. But he was unable to learn and profit from the experience — the telegraph called to him as nothing else had ever done — and soon he was off on his telegraphic travels again, fiddling with the equipment, making dumb jokes and dumber mistakes, and getting fired. In some ways, though not in all, he was a typical Knight of the Key, as his fellow telegraphers styled themselves, a group of men with strange lives, stranger talents, and a sometimes spectacular indifference to normality as it was practiced by the bulk of their fellow citizens.

One of Edison's friends, Milton Adams, went to San Francisco with sixty cents to his name, became a strikebreaker, sold patent medicine, and moved to Peru with the idea of setting up fights between a grizzly bear and a bull. The bear lasted five minutes. Adams then moved on to Buenos Aires, started a market report bureau, went broke again, and set up in the restaurant business in Pernambuco. But no place could hold him for long, and soon he fetched up in the Transvaal, running a panorama (a forerunner of the movies) called "Paradise Lost," which utterly failed to captivate the natives and separate them from their cash, whereupon he took work as a newspaper editor. Later still, he turned up on Edison's doorstep with a power of attorney, $2,000, and blueprints for a radical new invention that, it turned out, was roughly five thousand years old and in common use, although Edison neglected to say what it was. "The life has been somewhat variegated," Adams said, "but never dull."[2]

Edison, by contrast, was a dull boy indeed, although he was good company and a droll enough fellow. Still fascinated with machinery and working in Adrian, Michigan, southwest of Detroit, he hooked up two old Morse registers — the printing devices that embossed the message on a tape — and created a primitive repeater. This was a device that copied the message at a slower speed, which allowed Edison, with his limited skills, to transcribe it. In Louisville, one of the few new and modern offices he encountered on his odyssey, he unbolted all the equipment, the keys and the annunciators and at least some of the acid containers, fiddled with it, and dropped a container of battery acid that ate through the ceiling of the bank below and destroyed the rug and furniture.

"Edison traveled around the country," John Dos Passos wrote in the novel USA,

taking jobs and dropping them and moving on, reading all
the books he could lay hands on, whenever he read about
a scientific experiment he tried it out, whenever they left
him alone in a telegraph office he'd do tricks with the
wires . . . always broke, his clothes stained with chemicals,
always trying tricks with the telegraph . . . he worked all
day and all night tinkering with cogwheels and bits of
copper wire and chemicals in bottles, whenever he
thought of a device he tried it out . . . whenever he got a
hunch he tried it out.[3]

His sense of humor, such as it was, did not desert him. In Cincinnati, where he finally made first-grade operator, he hooked an induction coil up to a washbasin and looked on from hiding as his
fellow workmen danced comically about in the aftermath of the
shock. In Boston, his last posting as a telegrapher, he hooked up his
coil to the drinking pail.

But by 1867, in Boston, sleeping on the floor in Milt Adams'
rooming house, either he was ready for a change or a change found
him. Boston, as it happened, was the center of the country's electrical
engineering profession, which wasn't saying much. There were only
about two hundred trained engineers in the entire country, and only
a handful of them knew anything useful about electricity. In Boston,
those that did inevitably gravitated to the shop of Charles Williams,
who sold electrical components, encouraged his young clientele in
their experiments, and arranged for them to take laboratory space in
his building. In the window of his shop was an incandescent light
built by Moses Farmer, the foremost and possibly the least ambitious
electrical inventor in the country. Edison inevitably gravitated there.
And in Boston, for the first time in his life, he read Faraday's *Experimental Researches in Electricity*, a compilation of scholarly articles written by the great English scientist between 1839 and 1855.

Edison, so voluble on most subjects as he grew older, was curiously reticent about his personal life, and he did not confide in his
public about his ambitions during this period. Still, fate seems to

have taken a hand in guiding his steps. George Milliken, the manager of the Boston Western Union office, had taken a shine to young Edison and introduced him to Frank Pope. Pope, seven years older than Edison, was already the author of *Modern Practice of the Electric Telegraph*, a book destined to go through fifteen editions over the next twenty-three years. Pope was also a cofounder of the influential *Telegrapher* magazine, which soon announced that Edison planned to set himself up as an inventor. Specifically, Edison planned to invent an improved duplex telegraph, a method pioneered by Moses Farmer in 1856 for sending two messages simultaneously over a single wire, in effect doubling the size of the system.

But an improved duplex telegraph was not what Edison ended up inventing, at least not yet. Instead, for whatever reason, he ended up inventing an electrochemical vote-counting machine that used the principles of Faraday to record the conclusions of a deliberative body, such as the House of Representatives. For it, he received his first patent. Nobody wanted it. "Young man," said the chairman of the congressional committee to whom he brought his device, "that is just what we do *not* want." A machine that instantly recorded the legislators' votes, it was explained to the young inventor, would abolish the all-important deal making that occurred while the ballots were being cast, and deal making was the warp and weft of American democracy.

Next, with the backing of a small-time investor, Edison built a version of the telegraph-based stock ticker invented by E. A. Callahan in 1867, strung his wires over the roofs of Boston, and sold about thirty of them. Edison had developed a pattern: the duplex was invented by someone else but in need of improvement, the vote counter was based on Alexander Bain's telegraph invented in England, and the stock ticker had just been invented by another American. Edison took someone else's idea, rang a few changes on it, and made no money. Soon, leaving his backer behind, Edison, waiting for something to turn up, was sleeping on the floor of the battery room at the Gold and Stock Exchange in New York, where Pope was the supervising engineer. One day, something did. A spring fell into one of the cogs of the exchange's central gold reporter, another telegraph-based device that recorded the metal's current price. Pandemonium reigned in the marketplace, Edison fixed the machine, and his fortune was made. The young man who saved the day became the partner of the respected Frank Pope.

Like many men in his day and our own, Pope apparently enjoyed patronizing a bright lad who seemed to have some kind of a future, not least because Edison proved capable of doing all the work in their new joint venture, maintaining and improving the technology of the gold exchange. Edison's years on the road, fiddling with things and getting fired and moving on, had not been wasted. He knew a great deal about telegraph systems, and in 1870, almost everything in the world that was electrical was a child of the telegraph. With his connections to the gold exchange, he began to improve the design of the price reporters. He also went back to work on his planned improvement of the stock ticker. And there was his duplex telegraph. And there was Jay Gould.

Gould, building up to his celebrated but brief corner of gold (in 1869 he momentarily controlled its price), was extremely interested in the gold market. The man known to his contemporaries as the Exterminating Angel of Wall Street, he was also interested in the stock market. Indeed, anywhere a marketplace existed, where the price of something was set and could be manipulated to his advantage, was an interesting place to Jay Gould. Anyone interested in the stock market was by definition interested in the railroads, virtually all of whose securities were listed there (Gould had already, in another celebrated phrase of the period, reduced the Erie to two streaks of rust in Pennsylvania), and the railroads led his eye to the telegraph, whose wires ran beside the tracks and managed the traffic of the trains. Nothing was ever simple about Jay Gould. Railroads also (and immediately) led him to Commodore Cornelius Vanderbilt and his family, who controlled the New York Central and who, providentially for Gould and his profit, were definitely not in favor of Gould's control of the Erie, a competitor. The only way to stop him was to buy the Erie's stock, enabling Gould to make a pretty penny when he rigged the price in his favor. And the commodore also led Gould to the telegraph system: the Vanderbilts controlled Western Union, and another possibility arose to separate the old boy from some of his money. Gould was not simple and he was not a stupid man. He had perceived something that was not obvious to others. There was another person in whom all these things —gold market, stock market, railroads, and the telegraph — came to-gether: young Thomas Alva Edison.

Edison was more than reticent about his relations with Jay Gould. In his authorized biography, he lied about them. Perhaps, for his peace of mind, it was just as well. Gould, the personification of the piratical era in the nation's finances that Mark Twain named the Gilded Age, was not a good person to have known, especially if one had a reputation to consider. Edison's shadowy association with Gould took the science of electricity to a place, the corrupt side of the financial markets, where it had never gone before. Morse's orig-inal backers, the corrupt (but in another way) Fog Smith and the Vail family, had been active participants in his venture, but the number of Morse's backers was small and their original shares were closely held and not traded. It was only later, when Morse, his team, and their imitators had the telegraph up and running that they at-tracted the attention of the financial community. Field, a business-man, had attracted the attention of businessmen even wealthier than himself, largely in England, and the cable had been laid with off-the-shelf technology. Moreover, both Morse and Field had leaned hard on help from the government. This was not the case with Edison. His knowledge of capitalism was rudimentary and it never improved, and he did not batten at the government trough. Almost from his arrival in New York he found himself deeply in-volved with the men, like Jay Gould, his generation called the Money Interest. This involvement was twofold. First, there was the input of Edison. He did not understand capitalism and even his own personal finances were chronically in chaos, but he understood the technology of electricity and he had some inkling of what it could be made to do. Moreover, he was eager to make electricity do it, for which he needed a sum of money. Second, because he was so conspicuously knowledgeable about electricity, he succeeded in attracting the attention and the dollars of men who did not understand what he was doing, could not do it themselves, and hoped that Edison would somehow make them a lot of money. In the language of the flowers as spoken on Wall Street, electricity had become a speculative vehicle. Morse and Field had shown the way. And Edison, albeit briefly, was able to turn his inventions into money and cash in.

They were an odd pair, Edison and Gould, but the times themselves were odd. Gould, Edison said, was the most humorless man he'd ever met. Courting Edison, the financier invited him to his home and attempted to entertain him with talk about the railroads and the stock markets. Edison, who was interested in neither railroads nor stock markets except as they pertained to machinery and electricity, tried to keep the evening going by telling funny stories. Unsurprisingly, they did not exactly hit it off. But Gould, for reasons of his own, twice saved Edison from the consequences of his financial folly. Edison returned the favor by saving Gould's new telegraph company by inventing a new technology. And Edison, like Gould, had attracted the attention of Commodore Vanderbilt.

As Edison soon discovered, a man who dined at Commodore Vanderbilt's table was advised to sup with a long spoon. William Orton, the former salesman of religious books who headed Vanderbilt's Western Union, knew who buttered his bread. It was Cornelius Vanderbilt, and the commodore's instructions were clear: everything pertaining to Western Union was to be squeezed for every penny that could be yielded or trimmed. And when the ingenious young Thomas Edison attracted the attention of William Orton, it became clear that Orton had a fine instinctive mastery of the old bait-and-switch. Orton promised something that Edison wanted badly: money. Then, sensing that Edison had no understanding of either money or the laws of contract, Orton changed the rules in the commodore's favor.

First, the bait. There was no doubt that young Edison was good at what he did — so good, in fact, that Western Union soon bought out the Pope and Edison patents to the improved gold reporter. Edison's reward was $5,000 and a mandate to keep up the good work with a promise of further riches to come. In 1870, following a path that would inevitably add him to Vanderbilt's collection of trophies, Edison took the money and decamped to the industrial district of Newark, where he set out to make his fame and fortune as a sort of glorified machinist who could be called upon to solve the problems that had eluded solution by others. The gold reporter needed further improvements. So did the stock ticker. Edison delivered them, improving the commodore's prospects for future income. But the improved duplex telegraph, one of Edison's hardest and most enthusiastic projects, did not work. Orton found this useful. Edison had gotten away with five thousand dollars of Vanderbilt's money, but he would soon be back for more to finance his next invention. When he did, Orton could point to the failure of the duplex as a precedent for possible failures to come, and he could beat down Edison's price. The young man knew where to go for money, and he had gotten rather a lot of it. Now, because of the failure of the duplex, he was going to work on the cheap.

Edison developed an improved printing telegraph that worked at a blazing two hundred words a minute when an operator at the top of his game could send forty-five words a minute. He sold it to a new but poorer set of investors, but he also presented it to Western Union. Western Union didn't want it. For Orton's purposes, it helped that Edison, innocent in business and ignorant of the most rudimentary accounting practices, spent every penny he laid hands on. Because he knew something of the lad's financial plight, Orton's argument now went like this: If Edison came up with something Western Union actually needed, the company would buy it at a bargain price. If Edison didn't come up with something Western Union needed, well, he was not the only inventor in the world. Edison might be good in his own limited way, but he was not essential to the future prosperity of Orton's telegraph company. The next time he had an invention, he would, of course, improve his prospects by cutting his price. Deeply. Edison, Orton hoped, was now between a rock and a hard place. He would come around; most people did. But then a bug seemed to bite him.

Like other small industries of the period, Edison's new establishment in Newark was known as a German shop. He employed native-born Americans, a number of them, some of whom stayed with him for years, but he leaned heavily for his workforce on recent immigrants from Central Europe, skilled and industrious men like the literal-minded and gullible John Kruesi, who ran the machine shop and could build anything Edison could conceive. It was much less common to employ Englishmen. The class system had given them poor work habits, they had their empire as an outlet, and they tended to be adventurous, something that was not needed in a machine shop or around devices that employed electricity. But Edison hired one.

Charles Batchelor was two years older than Edison. He was the son of a mechanic, schooled in the heart of the Industrial Revolution in Manchester, and a former employee of the Coates Thread Company. He had an orderly mind and kept orderly notes. He was a skilled draftsman. He had imagination. He could be relied on to work alone. And he was intensely loyal. He was, in short, the ideal Edison employee. Edison, during all his years, even when he turned his attention to the afterlife, thought of the universe as organized like a factory. There were the workers, anonymous no matter what they actually did, and there was the leader. The leader got the credit. By temperament, Batchelor fell in perfectly with this scheme. In all but name, he was Edison's partner for many years. As the inventions began to flow, it was often difficult to tell where Edison left off and Batchelor picked up, and the inventions did not flow until Batchelor arrived on the scene. He died a rich man; Edison was generous. But almost no one has ever heard of him, because in Edison's world there was only one leader, and the leader got the credit.

Perhaps Batchelor was only a sidekick, but he kept Edison on a tight leash. Previously Edison had been all over the map, moving from project to unrelated project: the duplex, the voting machine, the gold register, the stock ticker, and back to the duplex again. There was a certain start-and-stop logic to all of this. Edison's mind always hopped around, which was a strength and also a weakness. Inspired by his reading or by something observed in an experiment, he would begin a new project, sometimes with startling results. But other times he would begin a new project, lose himself in it, let everything else go to hell, and waste a simply enormous amount of everyone's time. He was intensely, obsessively curious about the nature of things and what they could be made to do, and it was this obsessive curiosity that was the secret behind the phenomenon that became known as Edison. At Newark, with Batchelor at his side, there began to be a certain progression in his work, at least some of the time, as one thing began to suggest another. Serendipity — the finding of things by surprise — never ceased to play a role in his work, but a measure of unity also appeared.

The duplex had failed, and it didn't help matters that Edison had oversold it to Orton and Western Union. He always oversold the things he did. He was always about to roll something out in a few days or a few weeks and he always announced the completion of something before he was done. As a result, in the years before he invented the phonograph, Edison developed a somewhat justified reputation as a chronic liar. He also, for reasons that had largely to do with Jay Gould, developed a reputation as an untrustworthy sneak.

Failed or not, the duplex had revealed certain principles and possibilities. If you could reverse the polarity in the line — using the terms Franklin invented, change it from positive to negative and keep repeating the process — you could indeed send more than one message at a time, because you had, in effect, two or even more lines, depending on how fast you made the switch. And it got better. You could send messages in the opposite direction by increasing the voltage. And, Edison had discovered, the line's polarity could indeed be reversed. The duplex test — between New York and Rochester, the home cities of Western Union — must have failed for other reasons. Perhaps there had been electromagnetic interference, the bane of early telegraph experience. But if the line's polarity could be reversed — and Edison knew it could — why not send four messages? It would be quadruplex telegraphy.

It was, however, accompanied by a catch. As it became clear that Edison was actually going to achieve quadruplex telegraphy, multiplying the capacity of the existing lines by four, Orton insisted that Western Union's chief electrician appear on the patent as coinventor. Although Western Union, like Frank Pope, hadn't done a stitch of work, half of the royalties would now belong to the company. Although Edison had other customers, around the Newark shop and Edison's homestead it was Western Union that paid the bills, and there was now no Western Union money coming in even though Edison had assigned half the quadruplex patent to the company. Nor were there any savings, any cash reserve. Edison had a bad case of tunnel vision, and he was also — and always — an intensely self-centered man. He never learned to stop spending money as fast as it came in, either to buy new materials for his shop or to purchase lavish gifts for his new wife and former employee, Mary Stilwell, a young woman seventeen years his junior whom he otherwise largely ignored. Now, with the great depression of 1873 raging through the land, his creditors, never particularly tolerant, had become positively rabid. The sheriff was literally at his door. When Edison asked Orton for money, Orton said he'd think the matter over on his vacation. When the quadruplex was deployed, the annual savings to Western Union would be half a million dollars. Clearly, it was time to let Edison sweat a little.

It was also time for Jay Gould to take a hand, for a long-awaited fly had just fallen into his web. As Gould no doubt intended, the exact sequence of ensuing events is far from clear. (Interestingly, Gould may have been one of Edison's Newark landlords and another of his many creditors, further softening up the target.) The Automatic Telegraph Company, another of Edison's customers, advanced ten thousand dollars, perhaps on Gould's instructions. The money gave Edison a brief respite, while Western Union continued to waffle and delay about the quadruplex. There was still no deal in that direction. Seizing the moment, Gould bought Edison's half of the quadruplex patent for $30,000 — the price, Gould pointed out in what passed for him as a witticism, of *Plymouth Rock*, the yacht he'd just sold. Edison squandered most of the money. He also signed on as chief electrician of the Atlantic and Pacific Telegraph Company, Gould's own venture in the field.

The ensuing lawsuit between Gould and Western Union dragged on for a year, not a thing calculated to disturb Jay Gould. Paralyzing lawsuits were one of his favorite tactics, and he was aiming at larger game. He proposed to take over Western Union, and eventually he did. But not before his thirty-thousand-dollar investment in Thomas Alva Edison paid a rich dividend. "That young man," Orton remarked somewhat disingenuously of Edison to a British visitor, "has a vacuum where his conscience ought to be."[4]

* * *

Meanwhile, the young man with a vacuum for a conscience was inventing up a storm in Newark. Sporting the only moustache he ever wore — it resembled a smudge of dirt on his upper lip — and going gray before he reached the age of thirty, he invented an electric typewriter and wisely set it aside. Typewriter technology, which he never looked at again, was in its infancy. Previously he had used induction coils as a means of shocking his coworkers as he watched from a place of concealment, but now he marketed an induction coil as a cure for rheumatism, gout, sciatica, and nerves; it was also (the advertising implied none too subtly) a splendid practical joke. He cobbled together a patent medicine called Polyform. The principal ingredients were morphine, chloroform, ether, chloral, and spices; it was "for external use only," the bottle warned. He tried to invent a chimneyless kerosene lamp, a substitute for ebony made from sawdust, and a refillable cigar. He invented celluloid and lost interest in it. He was still all over the map, but a certain discipline was beginning to emerge. He invented an electric pen, a complicated device that was the predecessor to the mimeograph and proved to be his first successful creation not derived from the telegraph. The electric pen suggested a possible electric scissors, a dental drill, and an electric sewing machine. He set them aside and invented the mimeograph.

He also invented something called the electromotograph. With it, he saw the last of Jay Gould. There wasn't much lab space in the Ward Street shop in Newark, but he and Batchelor made the most of it. Experimenting with a penlike stylus and a cylinder wrapped in chemically treated paper, they noticed that the resistance of the paper to the stylus decreased when the stylus was electrified. What happened next sounds like a great deal of fooling around, which was, in fact, the secret of the Edison inventive method. Searching for an even more efficient coating on the paper, he and Batchelor — or "Batch," because Edison had to have a nickname for everyone — tried catnip, tellurium, dandelion, aluminum, chickweed, and a leftover breakfast of coffee, eggs, sugar, and milk. It's hard to take this sort of thing seriously. Edison never became the master chemist that legend depicts. Although he'd had a childhood laboratory in the basement of the home in Port Huron and he'd read a certain number of books on the subject, he had so little confidence in his own abilities that he took classes at Cooper Union. But the impulsive boy who couldn't resist playing with telegraph equipment became the impulsive man who couldn't resist playing with chemicals. Given a problem in an experiment, he would throw anything at it, sometimes in ways that didn't seem altogether sane. But the method worked, and for a very simple reason: given enough chemicals or other substances, there was a possibility he would find one that worked. In the case of the stylus-and-cylinder device, the substance that worked was chalk. In technical language, he had just created a differential friction device. In layman's terms, he had built a switch, a device for turning electricity on and off. He called it the electromotograph. And a moment came when Jay Gould badly needed it, because the patent for Edison's switch was not controlled by Western Union, and Jay Gould was running a telegraph company that needed switches.

Western Union controlled the rights to the Paige relay, a switch in all respects identical to the one Henry believed Morse had stolen in Europe. Gould controlled Atlantic and Pacific (A&P), which used Paige relays; it was impossible to run a telegraph system without them, because in the absence of Paige relays, a branch telegraph could not tap into the main wire. Western Union yanked A&P's license to use them, disaster loomed, Edison dramatically arrived on the scene with his electromotograph, and the day — and Jay Gould's bacon — was saved. Gould repaid the favor by swindling Edison out of a promised payment of $250,000. And so they were even, more or less, and Edison escaped with a whole hide, which was more than many a Gould associate did.

In the Gilded Age, such shenanigans were all in a day's work. Western Union found Edison too valuable to dispense with, the electric pen was selling well, and Edison was modestly flush for the first time in his life. He had also proved — and not only to himself — that he could sustain himself as an inventor rather than a manufacturer, although the manufacturing business would continue to call to him for the rest of his life.

From a contractor-machinist who made his living improving the inventions of others, he would become a contractor-inventor, creating entirely new things that would be financed by others, who would then share in the profits. Promising a new invention every couple of weeks and a "really big hit" every three months, he upped stakes in the centennial year of 1876 and moved out of Newark. He had in mind the creation of the first professional research and development lab in the world, and the New Jersey hamlet of Menlo Park, where he chose to settle with a handful of picked men from the Ward Street works, was to become America's first town devoted to science, if only because little else ever occurred in the place.

Menlo Park in Edison's day is often thought of as a remote farming community. There were a handful of houses, an inn, a bar, and soon a boarding house run by Sarah Jordan, one of Edison's remote relatives, where a number of his men were put up. In fact, Menlo Park was a kind of protosuburb. Newark was only a few miles away, the town was on the main line of the Pennsylvania Rail Road, and the towers of the rising Great East River Bridge — the Brooklyn

Bridge — were visible by telescope from the front porch of Edison's spacious new house. If the need called — and it often did — he could get to New York, conduct business, and return by ferry and train the same day. The new lab was not, by modern standards, an impressive sight. Built by Edison's father, Sam, and called the Tabernacle, it was a two-story frame structure of a hundred by thirty feet. Within, there was a machine shop. There were more than 2,500 bottles of chemicals lined up on shelves that sometimes crossed the windows. They were on hand in case another fit struck Edison and he decided to solve a problem by throwing things at it.

The first big project was the telephone. Edison, working on yet another new form of telegraphy for Western Union, had almost invented the telephone but didn't realize it until later. Instead it was Alexander Graham Bell, the Scottish-born teacher of the deaf who never invented anything else worth speaking of, who demonstrated the first working model at the Philadelphia Centennial Exposition in the same year Edison moved to Menlo Park. Using a knowledge of carbon derived from his work on the automatic telegraph and responding to a Western Union request for a telephone that would smash the upstart Bell to flinders, Edison devised the carbon transmitter that turned the telephone from a toy that transmitted for yards into a commercial product that transmitted for miles. The Edison transmitter was part of every telephone handset for more than a hundred years, because Edison had found that carbon creates electrical waves that are analogous to the waves of the human voice, no other substance answered so readily, and it was a low-power device that used the telephone's trickle of electricity with maximum efficiency. Next, drawing on his experiences with the electromotograph, he devised an improved chalk button receiver that left something to be desired. For one thing, it was actually a cylinder of chalk, and to work it had to be cranked by hand. For another, it was a deaf man's invention. Reception was so loud that the telephone could be heard all over the house. Then, with his new transmitters and receivers, he set out to conquer the British market for himself, leaving America to his old client, Western Union.

The comments of George Bernard Shaw, who worked for Edison's British operation, are instructive:

> Whilst the Edison Telephone Company lasted, it crowded the basement of a high pile of offices in the Queen Victoria Street with American artificers. These deluded and romantic men gave me a glimpse of the skilled proletariat of the United States; and their language was frightful even to an Irishman. They worked with a ferocious energy which was out of all proportion to the result achieved. Indomitably resolved to assert their republican manhood by taking no orders from a tall-hatted Englishman, whose stiff politeness covered his conviction that they were, relative to himself, common persons, they insisted on being slave-driven with genuine American oaths by a genuine free and equal American foreman. They utterly despised the artfully slow British workman who did as little for his wages as he possibly could; never hurried himself; and had a deep reverence for anyone whose pocket could be tapped by respectful behavior. Need I add that they were contemptuously wondered at by this same British workman as a parcel of outlandish boys, who sweated themselves for their employer's benefit instead of looking after their own interests. They adored Mr. Edison as the greatest man of all time in every possible department of science, art and philosophy, and execrated Mr. Graham Bell, the inventor of the rival telephone, as his Satanic adversary. They were free-soul creatures, excellent company; sensitive, cheerful, and profane; liars, braggarts, and hustlers; with an air of making slow old England hum which never left them even when, as often happened, they were wrestling with difficulties of their own making; or struggling in nothoroughfares from which they had to be retrieved like stranded sheep by Englishmen without imagination to go wrong.[5]

In Menlo Park Edison continued to sleep on tables and floors, he invariably sported a huge chaw of tobacco in his cheek, and his only concession to the weakness of the flesh was a regular midnight lunch served to his crew, sometimes followed by a sing-along to an organ donated by Theodore Roosevelt's cousin Hilborne Roosevelt, the country's foremost manufacturer of the instruments. There were never any clocks in the invention factory, and Edison never knew and never cared what time it was. Once, belatedly realizing that a new hire hadn't secured lodgings, he told him to take the rest of the day off and find a room. The new man looked out the window. It was the middle of the night. And, in the middle of the night, his men would sometimes hear Edison's measured stride on the boardwalk he had built as, hands in trouser pockets according the fashion of the day and with a wide-awake or a skullcap on his head, he walked home to spend a rare night in a bed.

But the deliberately fostered notion that Menlo Park was a place of ceaseless if happy toil was a false one. Edison was still running a German machine shop. The celebrated all-nighters occurred, but they occurred only when the Old Man was in a hurry or deeply interested in something. Afterward, the men would return to the boarding house and crash until the following afternoon. The work schedule was set by what Edison wanted to do, or by what he was physically capable of doing. An early employee, Francis Jehl, reported that Edison once slept for a solid thirty-two hours with a short break to eat a steak. And the work, by modern or even contemporary European standards, was often of a makeshift and primitive sort. Behind the Tabernacle was the lamp house, where dozens of kerosene lamps burned constantly and smoked up their glass chimneys. Every once in a while, someone came in to scrape off the carbon, which was then shaped into telephone receivers. Batchelor was especially good at it, and Edison would never again be so happy or so well adjusted to the reality of his surroundings. Years later, when he employed university-trained scientists in his vast new research facility in West Orange, some of them would deliberately smudge their faces with grime to show the boss their old-fashioned

zeal for the work although there was, in fact, no dirt at all in the an-
tiseptic labs where they labored and Edison should have known so.
The world changed, but Edison's view of it didn't. Somehow he always seemed to be back in 1876. As the world soon discovered, the
great inventor had a remarkably rigid mind.

Edison had never been so organized, and he would never be so
organized again. Mostly, the work revolved around the telephone.
Determined to dominate the new industry as it dominated the telegraph, although in the end it failed, Western Union was eager for
anything Edison could discover or invent. And Edison — exploring
the telephone, a device that reproduced the human voice, and on
the roll of his life — noticed something.

It is easy to forget how compressed the time was in Edison's great
days — that is, how much he did in so little of it. Only a few years
before the move to Menlo Park, as a wandering young telegrapher,
he had hooked two old Morse registers together and made a primitive repeater that transcribed the incoming messages so slowly that
he was able to write them down. Perhaps he could make a repeater
again, but in an entirely new way, on a paper disk or strip or — drawing on the electromotograph and the telephone chalk receiver — on
a cylinder, where the signal would be indented by a stylus rather
than printed. He already knew that sound could be turned into vibrations and vibrations back into sound; his carbon telephone transmitter did just that. "If the indentations on paper could be made to
give forth again the click of the instrument," he wrote, "why could
not the vibrations of a diaphragm be recorded and similarly reproduced? I rigged up an instrument hastily, and pulled a strip of paper
through it, at the same time shouting 'Halloo!' Then the paper was
pulled through again, my friend Batchelor and I listening breath-
lessly. We heard a distinct sound, which strong imagination might
have translated into the original 'Halloo!' " It was enough, he said, to
make him begin an entirely new line of inquiry.

Working all through the summer and fall of 1877, he and Batchelor concluded that the new device would be mechanical rather
than electrical. They experimented with waxed paper, ticker tape,

and chemically treated paper. The word "phonograph" began to appear in their notes. There was a premature announcement of complete success in the *Scientific American*. And finally, there came the night of November 29. Batchelor bet him a barrel of apples that it wouldn't work.

In Edison's version of the story, carefully crafted for popular consumption, it was a little paper cutout of a man sawing wood that gave him the idea. He sensed that most people either believed or liked to believe that inventive genius works in flashes, and so he gave them flashes. In the flash he contrived, he created his own version of a philosophical toy, and the toy gave him the idea. Or so the story went.

It had been known for more than fifty years that sounds, including the sound of the human voice, traveled in waves. Erasmus Darwin claimed to have invented a compressed-air device that reproduced human speech, but an assembled contraption wasn't found among his effects. It was duplicated later — well into the twentieth century — from his designs. Human speech, as emitted from Dr. Darwin's reconstructed apparatus, sounded something like a belch. In France in 1857 an inventor named Leon Scott de Martinville had found a way to record sound waves with a stylus on a lampblacked piece of paper, although he could not play them back. According to one story, he had a copy of Lincoln's voice, but the paper, if it ever existed, has been lost. Another Frenchman named Charles Cros had delivered a lecture in 1877 in which he described a reproduction device, but he killed himself with absinthe before he got around to building it. Edison, in a playful mood after designing the carbon transmitter for Bell's new telephone, had hooked up a diaphragm to a small funnel, placed it atop his desk, and attached it to the paper cutout. If he spoke loudly into the funnel, the arm holding the saw began to move rhythmically. His daughter Marion was delighted with it. His two children, a boy and a girl, were called Dot and Dash by his workmen.

It was the way Edison liked to work. After the early fiasco with his vote-counting machine, he proclaimed that he would never invent anything that did not have a clear commercial market. That said, if he got interested in something, no matter how trivial or strange, he would often pursue it and think up a justification later. In 1877, the year when he allegedly paused in his labors to contemplate the figure of the little sawing man, he was not yet famous. He wasn't even well known outside a small circle that included the Western Union Company and Jay Gould. All that was now going to change.

At the age of thirty, he was set in his ways. He was still a man who liked to work through the night, curling up on a worktable or lying down in a nice warm pool of acid whenever he needed a nap. As he grew older, he preferred to sleep in a rolltop desk, although a closet would do at a pinch, and among the many claims for himself was the boast that he needed no rest or very little. "Everything which decreases the sum total of a man's sleep increases the sum total of a man's capabilities," he said. "There really is no reason why men should go to bed at all." In fact, he slept about as much as anyone did, although only occasionally during the hours others reserved for the activity. "Look at this son of a bitch," said one of his workmen, pointing to a snoozing Edison in a chair. "He tells the world he never sleeps, but he is fast asleep like this pretty near all the time. He just don't believe in nobody else sleeping!" At the end of a marathon session in the workshop (and there were many), he would wander home in clothing that was more often filthy than not and collapse on a nice clean bed. Edison believed that removing his clothes caused insomnia.

No man liked a joke better, although his sense of humor was not universally shared. Assistants who fell by the wayside during one of his all-nighters were roused to renewed action by his "corpse reviver," a handheld electrical device that gave an invigorating shock, or were awakened by his "calmer," a noisemaker placed an inch from the ear. In Boston, on the last telegrapher's job he ever held, he narrowly escaped being brained with a glass insulator thrown by a superior enraged by a practical joke. His admirers, making a virtue of necessity, said that the boy was ever present in the man. "He had," said one, "the undimmed hope and joy of a guileless child."

"There is a general appearance of youth about his face," wrote a reporter with a somewhat clearer eye, "but it is knit into anxious wrinkles, and seems old. The hair, beginning to be touched with gray, falls over his forehead in a mop. The hands are stained with acid, his clothing is 'ready-made.' He has the air of a mechanic, or, with his peculiar pallor, of a night-printer. When he looks up his attention comes back slowly as if it had been a long way off. But it comes back fully and cordially. A cheerful smile chases away the grave and somewhat weary look. He seems almost a big, careless schoolboy."

And now, in his new Menlo Park workshop, the big, careless schoolboy with the nightworker's pallor was about to do something he had never done before and would never do again. He invented something from scratch.

Edison took a sheet of paper, made a rough sketch of a device, and wrote the figure $18. He handed the drawing to John Kruesi, the Swiss machinist he always called Honest John, and asked him to produce the device within thirty-six hours. If Kruesi succeeded, the bonus would be his. If he exceeded the time limit, he would receive his normal wage. Kruesi asked him what the device was supposed to do. Edison told him. Kruesi said it was impossible. Then he went off to earn his bonus. Everything Edison and his muckers had built to date revolved around electricity. This one didn't.

The device was a simple one; Kruesi easily earned his eighteen dollars. There was a cylinder wrapped in tinfoil, a crank, a base, a needle on an arm, and a funnel like the one that propelled the little sawing man. Edison turned the crank and recited "Mary Had a Little Lamb." Then he reset the needle and turned the crank again. His voice emerged — poorly but recognizably — from the funnel.

"Gott in Himmel," said Kruesi.

Edison himself was amazed.

Some of the story was even true. He left out Batchelor's part, of course, but that was normal. And unless pressed — or in another of his moods — he left out the months of labor.

* * *

In addition to the Old Man, they also called him the Beast. If he bathed more than once a week, it was a rare event, and he was addicted to chewing tobacco and pie. "The English," he'd said a few years earlier, "are not an inventive people; they don't eat enough pie." Later in life, he expressed his approval of President Warren G. Harding because the chief executive was fond of a chaw. He never lost either his Midwestern, midcentury accent or his Midwestern, midcentury prejudices. He said "somp'n" for "something," and "figgerin'" for "figuring." He was fond of coon jokes and suspicious of Jews. When he finally got his phonograph business up and running, he insisted on selecting the music for its cylindrical records, although he was extremely hard of hearing and had to bite a piano in order to hear it. When he went into the movie business, inventing his own camera, film, and arcade viewing device but buying somebody else's projector and putting his own name on it, his wholesome films were greeted with enthusiasm by the many censorship boards that had been set up across the width and breadth of the land. One of his films was the first *Frankenstein*. His largest undertaking was a huge Boob McNutt–like complex that mined iron ore that nobody wanted from the played-out hills of northern New Jersey, where the Edison-designed buildings sometimes fell down. Edison was no architect, but by then, like many powerful and famous men, he had begun to confuse himself with a force of nature. He lost a fortune in the iron business. Next, with salvaged machinery, he went into the concrete business and lost another three million dollars. In monetary terms, his most successful invention was the alkaline storage battery. He believed the automobile was an interim technology and the home radio was a fad. He once employed an executive who never made a decision without consulting, through a medium, a dead Indian named Old John. For years, Edison himself patronized a mindreader called Professor Bert Reese, who also claimed to have X-ray vision. To see if he could duplicate these advanced powers, as he called them, Edison once fitted himself and two of his friends with electrical coils, which they wore on their heads. Although he later denied it, he was an early member of the Theosophical Society, whose founder, Madame Blavatsky, referred to him as Brother Edison, and he kept a large number of spiritualist works in his library, including *The Evolution of the Universe*, a book supposedly dictated by the late Michael Faraday in 1924.

The 1877 phonograph was not a commercial product, although until then Edison had announced nothing that was not a commercial product. It was a prototype, it had great potential, and people were fascinated with it. But it wasn't ready for the big time, not yet. The recording medium was tinfoil that had to be carefully wrapped around the cylindrical receiver. Sometimes it tore. Even under the best of circumstances, a steady hand was required on the crank that turned the foil-wrapped cylinder, either to record or to broadcast. To achieve anything like verisimilitude, the cylinder had to be cranked at the same speed when recording and playing, no easy trick. The foil lasted for only a few replays. There was a funnel that one spoke into and also listened to, and to achieve a proper recording, it was required that the recording party bellow at close to the top of the lungs. The reproduction resembled, but was not identical to, human speech. If everything worked perfectly, it could be understood. And the world of 1877 went wild over the phonograph.

The next day, December 7, after Edison and the boys had stayed up all night playing with the new machine, Edison took his phonograph to the offices of his favorite magazine, the *Scientific American*. According to Edison, who may have been telling another stretcher, chief editor Alfred Beach had to clear the room, because he was afraid the floor would collapse under the combined weight of the spectators.

"Mr. Thomas A. Edison," the magazine reported in its next publication, "recently came into this office, placed a little machine on our desk, turned a crank, and the machine inquired as to our health, informed us that it was very well, and bid us a cordial good night."

"Speech," declared the editorial in the same issue, "has now become immortal."

Further exhibitions were held in Menlo Park. So many people flocked to see them that the Pennsylvania Rail Road laid on special trains.

Edison was summoned to Washington. By then, he had added a third tinfoil cylinder to his repertoire, a recitation of

> There was a little girl, who had a little curl
> Right in the middle of her forehead;
> And when she was good, she was very, very good,
> And when she was bad she was horrid.

Senator Roscoe Conkling, who had a little curl right in the middle of his forehead, was not amused. Perhaps he wasn't supposed to be. Senator Conkling, as counsel for Western Union, had put Edison through a particularly galling examination in the quadruplex lawsuit with Gould.

Edison met with the aged Joseph Henry at the Smithsonian. His picture was taken by Matthew Brady. Playing with a rubber band but not speaking (he had a horror of public speaking), he sat on the stage at the American Academy of Sciences as the phonograph told the assembled dignitaries how glad it was to be there. At eleven that night, a summons came from the White House. President Hayes and his cabinet were entranced. At twelve-thirty, the ladies were summoned. The demonstration lasted until three-thirty in the morning.

Not everybody was impressed. There are always sourpusses in the world. Presented with an account of the phonograph in the *New York Sun*, a Yale professor characterized the reporter as "a common penny a liner in the incipient stages of delirium tremens. The idea of a talking machine is ridiculous." Bishop John Vincent, the celebrated Methodist churchman and cofounder of the Chautauqua gathering, was convinced there was a ventriloquist concealed somewhere in the room. Given a private audience at Menlo Park, he began to bellow into the funnel. "Mahalaleel!" he roared in his best pulpit manner. "Mathuselah! Aphaxad! Hazarmaveth! Chedorlaomer!" The list of names, Edison remarked, "could have stopped a clock." When the cylinder was replayed, the bishop was converted. No one on earth, he confessed, could have duplicated his diction.

The cylinders played for two minutes, but people couldn't get enough of them. Edison was particularly pleased that the exhibition in Boston was pulling in $1,800 a week. Of all his inventions, he was always proudest of the phonograph. It made him famous; his name, suddenly a household word, became synonymous with "inventor." Overnight, the young man who was known to the public (if he was known at all) as Jay Gould's sidekick became the Wizard of Menlo Park.

* * *

In the unknown country of the past, there had never been a machine that talked. The talkative Yale professor was not an entirely foolish man; such a thing was all but inconceivable. True, it was an age of invention, the swiftest and most fruitful the world has ever known, when a man like Lord Kelvin, the former William Thomson of Atlantic cable fame, could be born with the steam locomotive and go out, after a long life, with the airplane. But the transformative inventions of the nineteenth century were not merely understandable by analogy. It was no accident that the principal unit of measure for force was the horsepower. As mentioned earlier, for thousands of years, humankind had precisely three kinds of power at its command: muscle, wind, and moving water. These sources of power were now in the process of being replaced by steam. It was an extremely untidy event, a major and disruptive break in the flow of history, and at first it was not kind to most of the people — the workforce — it affected. Steam power polluted the air, fouled the waters, and turned the urban day into a kind of twilight. It confused people, and it angered them. They had to go to work at a certain time, they had to work throughout their entire shift, and they could go home only when the foreman said so. Almost nobody on earth had ever worked like that. Outside of a few industrial centers, the only organized body of people on the planet had been in armies and navies, and the people in armies and navies had to be beaten regularly. Indeed, that was one of the reasons the factory owners liked to hire lots of children, because children, like soldiers, could be beaten. The owners also hired women, because women were, taken as a whole, docile, and they were physically weaker. The owners, at least at first, did not hire many men. And the men deeply resented the Industrial Revolution. One of the first fruits of the Industrial Revolution was the English penal colony in Australia, a continent-sized prison that was soon inhabited by the former cottage weavers and nail makers who lost their jobs to steam, could not or would not find new ones, and took to crime. A goodly number of men, and not a few women, were broken on the wheel of the Industrial Revolution; wherever it went, predictions of revolution and social collapse followed, blood in the street and dirt in the sky and a vanished Eden forever receding in the distance, forever beyond the grasp of a weakening hand. And the Industrial Revolution, at least in its early phases, did not make a great many things that the common people wanted or could afford.

By contrast, Edison's talking machine, mechanical in action but electrical by derivation, did something that no amount of muscle power could accomplish. It sat on a table and spoke. The phonograph and the cascade of purely electrical devices that soon followed it promised comfort, entertainment, and leisure — and, to a very large extent, they delivered on the promise. The age thought of electrical devices as benevolent, delicate, and friendly, achieving, in the words of one scholar, "spectacular ends through inconspicuous means." It is, perhaps, no wonder that the members of the press and the chattering classes, contemplating Edison's talking machine, the first of these devices, appeared to become slightly unhinged.

A much-quoted passage in the *New York Herald* is an example, remarkable only for its restraint. "Invisible agencies are at his beck and call. He dwells in a cave, and around it are skull and skeletons, and strange phials filled with mystic fluid whereof he gives the inquirer to drink. He has a furnace and a cauldron and above him as he sits swings a quaint old silver lamp that lights up the long white hair and beard, the deep lined inscrutable face of the wizard, but shines strongest on the pages of a huge volume written in cabalistic characters. . . . The furnace glows and small eerie spirits dance among the flames."

The "simple inhabitants" of Menlo Park, it was elsewhere remarked, "were minded of the doings of the powers of darkness." On April 1, the *New York Daily Graphic* ran the headline "Edison Invents a Machine that will Feed the Human Race: manufacturing Biscuits, Meat, Vegetables and Wine out of Air, Water, and Common Earth." Papers all over the world, failing to take note of the date, copied. As late as 1890, the *Catholic World* ran a mathematical analysis of Edison's name and proved that he was Satan.

"Had the skies been suddenly darkened by a flotilla of airships," said one of Edison's lab assistants, "[bringing] a deputation of Martians, the phenomenon would have been accepted as a proper achievement of the scientist's genius." On a more secular level, it was suggested that the new Statue of Liberty, then rising in New York harbor, be equipped with a gigantic phonograph that would greet visitors and immigrants and broadcast uplifting messages to the inhabitants of the city.

Meanwhile Edison himself carried on, basking in the glow of an astonished nation's praise. He discovered that he loved to show off to the adoring public that flocked to Menlo Park for demonstrations of his marvel. Turning the crank for the assembled multitudes, he recited poetry, scripture, Shakespeare, and doggerel, happily interspersing the lines with cries of "Help, Police! Help!" and "Shut Up!" Then he played it back. The act laid them in the aisles.

Still, from the invention of the phonograph to the end of his life, it was believed he could do wonders. In 1892 a rumor appeared in France that he was building the Kaiser an "infernal machine" that would destroy a city at thirty miles and annihilate an army corps. He was the protagonist of Garrett Serviss's 1898 novel *Edison's Conquest of Mars*, where his character invented the space suit and the death ray. When the U.S. entered World War I, it was widely expected that he would end the war with a wonder weapon.

In fact, the only weapon he ever invented was a wire-guided torpedo that didn't work very well, and he didn't invent all of it. For the first time in his life, he was a member of a government committee, and the rest of the committee pitched in. His principal contribution to the war effort was a sea anchor designed to turn a ship in its tracks, enabling it to evade a U-boat attack; when deployed in practice, it stopped the ship dead in the water, making it a sitting duck. In the French novel *L'Eve Future*, he lit his cigar with an artificial camellia. For years, with Edison doing his best to add to the confusion, it was extremely hard to think straight about Thomas Alva Edison. For six years, between 1876 and 1882, he owned the world.

But Edison was entirely serious about the phonograph. It was not a toy, he told his new private secretary, Alfred Tate. In a list of possible uses, he included the recording of the last words of the dying — a favorite preoccupation of the times — and the preservation of the voices of the great deceased, but entertainment did not stand high. Instead, he believed he had invented the paperless office. Then he set the phonograph aside for ten years.

Something else, as often happened, had engaged his attention. As almost always happened, the idea for another invention had come to him — had come to him, in fact, in a flash. Demand for the phonograph was immense, and at first he had believed he was just the man to improve the thing and take the world by storm. He went so far as to set up a small manufacturing operation that produced replicas of the original crude machines for a hundred dollars each. Then, seemingly out of the blue, he sold the patent to a partnership of Alexander Graham Bell's father-in-law and cousin Chichester Bell. He was off in a different direction altogether. He had decided to create the world's first electrical power station.

CHAPTER SEVEN

LIGHT

If Edison had a needle to find in a haystack, he would proceed at once with the diligence of the bee to examine straw after straw until he found the object of his search. . . . I was a sorry witness of such doings, knowing that a little theory and calculation would have saved him ninety per cent of his labor.

— *Nikola Tesla*
October 19, 1931

Brockton, Massachusetts, summer 1884. Outside Edison's new central power station, a lightning storm raged. Inside Edison's new central power station, a lightning storm also raged. Naked wires arced, shooting off brief, bright, writhing, and crackling curves of visible electricity like something in Dr. Frankenstein's laboratory, that touched down on other naked wires, on the dynamos, on anything in the room that was made of metal. Even more spectacularly, the dynamos also began to arc. Then they began to "hunt," a commonplace term for a frightening phenomenon: the dynamos no longer quietly hummed away at a steady level but started to generate power surges. "Hunting" made it seem like the place was about to blow up, melt down, or shake itself to pieces as the entire structure began to vibrate. The caretaker, a former locomotive fireman chosen for his steeliness of nerve and familiarity with large explosive objects, ran for his life and was never seen again. Properly insulating his equipment was not Edison's strong suit.

* * *

Almost eighty years after the first Volta pile, most of the world's electricity still came from batteries. If Edison actually got around to inventing the devices that swarmed through his head — the electric dentist's drill, the electric sewing machine, the electric scissors, although he never did, in fact, get around to them — they would still require batteries consisting of strips of metal in a bath of acid. Edison had personal experience in Louisville, where he had upset the telegraph batteries, with what could go wrong. Moreover, it was no way to create a mass market. The electric pen, powered by its own small battery, had been his most successful private (as distinct from corporate, as in Western Union) invention, but only forty thousand of them were sold. An electric phonograph — he had already conceived of one — would probably do somewhat better in the marketplace, perhaps dramatically better, but it would be a clumsy device with its own dangerous acid battery, and it would require constant maintenance. There had to be a better way to generate electricity, and there was.

Faraday had long ago demonstrated that electrical power could be produced by mechanical means, and crude dynamos, the nineteenth-century term for a machine that generated electricity, had begun to appear in the United States and Europe. All you had to do was run an electric motor backwards and you had a dynamo, and a dynamo did not perform work the way a steam engine performed work, by moving a wheel that either propelled a locomotive or provided the visible mechanical power that ran the machines in a factory. Instead, a dynamo produced (usually, if things were working properly) invisible electricity that ran through wires to an electrical machine such as a lathe, and the electrical lathe performed the work. The connection between a steam engine, with its mighty chuffing, great moving pistons and rotating wheels, and the tasks linked to it were clear. The connection between a humming electric dynamo and an electric machine was not, because the exterior of the dynamo had no moving parts. But unless it had a battery, an electric lathe could not work without a dynamo.

Dynamo theory hadn't advanced much since the 1830s when Faraday built the first one as another of his philosophical toys. It was known that a dynamo generated electricity by moving a bar of soft iron, known as the armature, through a magnetic field created by surroundings of wire, but not much more than this was known. The winding of the electromagnet that created the necessary magnetic field and the management of the heat produced by the resulting electricity were up to the individual inventor. Building a new kind of dynamo was a hit-or-miss proposition. Sometimes the prototype worked and sometimes it didn't, and although some successful inventors shared their secrets, others were not as forthcoming; creating a new type of dynamo was more an art than a science. Still, successful dynamos were now being used to power Humphry Davy's arc lights, the intensely brilliant form of illumination that appeared as a permanent, stable arc of electricity between two rods of carbon, which were gradually consumed. And arc lighting connected to a dynamo was proving to be immensely popular, because the dynamo solved the problem that had bedeviled electricity ever since Volta unveiled his battery. A battery could power a single arc light, and not for very long. A dynamo could power a number of arc lights, and they would continue to provide illumination until the dynamo shut down. The low-power problem that had plagued inventors for decades was on its way to being solved. Thanks to the dynamo, a man in the arc lighting business could get rich. Equally important, a man's investors could get rich. Edison's experiences in the economics of the Gilded Age had taught him a number of wrong lessons, most notably that the road to riches was paved with the supine bodies of one's enemies, but it had also taught him the uses of other people's money. And if he went into the lighting business, he would need quite a lot of other people's money.

The time, however, was ripe. In 1866, almost simultaneously, Werner von Siemens in Berlin, Wheatstone and Samuel Alfred

Varley in England, and Moses Farmer in Connecticut announced
the discovery of the embryonic but eminently practical "self excited" electrical machine — the forerunner of a dynamo that could
have commercial applications. Farmer, the coinventor of the electric fire alarm, who worked in his day job making torpedoes for the
navy and did not make good use of his time, finally got around to
exhibiting his self-exciting machine ten years later in Philadelphia.
But there was a problem. The dynamo and its forerunners generated
electricity, and the electricity could be used to power an arc lamp,
but there the march of science halted. The electricity could not run
a useful electric motor, because there wasn't enough of it. It was the
old electrical paradox. You could run an impractical motor with a
battery. You could run an impractical motor with an 1870s dynamo.
Thomas Davenport's bane, the power shortage, persisted. Before
electricity could perform work as well as provide scientists with a
tempting phenomenon, an accident had to occur.

In Vienna in 1873, several of the continuous-current dynamos
invented by the Belgian Zénobe Théophile Gramme were being assembled on the floor of an exhibition hall. A careless workman accidentally attached a dormant Gramme machine to a pair of wires
originating in a Gramme machine that was already running. The
dormant machine immediately began to run backwards. It is probably due to someone's quick thinking that nothing further was done
until Gramme himself could be summoned to the scene.

Gramme immediately recognized what was happening. The formerly dormant machine was running like an electric motor. This
was the elusive key that everyone had been looking for: a dynamo
run backwards with the weak power of another dynamo became a
motor. The shape of the future was clear. Whenever anyone improved a dynamo, he would also know how to build an improved
electric motor. When the Scottish electrical scientist James Clerk
Maxwell was asked to identify the greatest discovery of the age, he
unhesitatingly replied: when Gramme's machine ran backwards.
Edison, when he turned his mind to the problem after setting aside

the phonograph in the later 1870s, was quick to sense the resulting opportunity. Creating a central power station was all very well as an abstract exercise, but it wasn't going to make anybody rich; it was, in fact, an invitation to throw money down a rathole. To actually sell the power, Edison needed electrical devices — lots of electrical devices that could be hooked up instantly and that people would badly want. The electric motor was one. Edison foresaw a rich market for them in the urban slums, where the country's industry was concentrated.

There was also reason to believe that there would be a rich market for electricity to power the arc light, a formerly dormant technology that was more than half a century old but was now showing dramatic signs of life, thanks to the new dynamos. In 1877, the same year that Edison turned his thoughts toward the central power station, Paul Jablochkov, a former Russian officer of engineers, introduced arc lighting to the city of Paris, illuminating half a mile of the Avenue de l'Opéra using Gramme dynamos. Paris was on its way to becoming the City of Light. In America, Charles Brush of Cleveland, Ohio, was another coming man in the arc lighting line.

As a boy, Charles Brush was far more inclined to ponder the discoveries of Humphry Davy than he was to do the chores on his parents' Walnut Hills Farm. He trained as a mining engineer and went into the Great Lakes coal and iron business, but somehow he could never shake himself free from the first enthusiasm of his youth; Davy's arc light still called to him. When word of the new self-exciting machines of the Europeans reached America, he realized that his chance had finally come. The Volta battery occupied the same premises as the devices it powered. So, it was assumed, would the dynamo, once it was perfected. Brush thought differently. Like Edison, and almost at the same time, he conceived the central power station, a location where, using dynamos, electricity would be generated.

In 1876, as a young man of twenty-seven who made his living selling pig iron, he built his first dynamo in his parents' barn. In

1878 he installed his first arc light on the Cincinnati porch of Alice Roosevelt's future father-in-law, Dr. Longworth, and he circulated in the crowd to sample its reaction. "In the gathering," he wrote,

> there were a few men of the type who, if they know nothing about a subject and find others who know nothing about it, will promptly explain it. One man who had collected a considerable audience called attention to the solenoid at the top of the lamp and said, "That is the can that holds the oil," and, referring to the side rod, said, "That is the tube which conducts the oil from the can to the burner." He said nothing at all about electricity — a little oversight apparently unnoticed by his hearers — and they went away happy in their newly acquired knowledge of the electric light.[1]

That same year, 1878, Brush lighted Wanamaker's department store in Philadelphia, and in 1879 he achieved his dream of a central power station when he illuminated a dozen street lights set atop fifteen-foot poles in his hometown. The Cleveland Grays band burst into music. Salutes were fired along the lakefront. Charles Brush built a mansion that was exceeded in splendor only by the local home of John D. Rockefeller.

But electricity, he found, remained an enigma to the mass of American humanity. Once, Brush, by then a captain of industry, traveled fifteen hundred miles to remove a tack that had shorted the coils of an electromagnet. At times workmen, irrationally resentful of the mysterious force or perhaps merely in a prankish mood, also used nails to spoil the workings of the electricity. Moreover, they wouldn't stop disassembling his apparatus and tinkering with it, even though they had no idea how it worked.

Brush devised new lamps that couldn't be taken apart. He generated electricity for his own house with an immense windmill in the backyard, retired from business when his company began a series of mergers that eventually made it part of General Electric, and spent the remainder of his life trying to devise an antigravity machine. After all, electricity had been a mysterious force and so was gravity. Brush was trying to get in on the ground floor again.

* * *

Edison became reasonably certain he could invent an electric motor — there is no indication he had ever heard of Thomas Davenport, the Vermont blacksmith who invented the first large one — and he decided to look into this new business of the arc light, something that hadn't previously attracted his attention. Clearly, it was becoming something of a rage. Edison's mind was ranging broadly then, although, for once, within a disciplined compass, and telegraphy was beginning to lose its charm. Western Union, for example, refused to pay him for his revolutionary carbon telephone transmitter until Western Union's president Orton dropped dead of a heart attack, whereupon the company finally coughed up the cash. And in more ways than one, the phonograph — a mechanical device based on electrical principles — seemed to liberate his mind. He was a scientific celebrity now, much as Morse and Field had been scientific celebrities respectively forty and thirty years before — so much so that he briefly staked a trained bear outside the lab in Menlo Park to keep the increasingly distracting multitudes away. He was not, however, a scientist, which poses the not-so-obvious question of just what, exactly, Edison was.

In later years when, like many extremely successful men, he came increasingly to confuse his thoughts with the laws of nature, he frequently invoked the name of science to explain what he did. It was noted, however, that the older he grew, the less he actually knew about science, in part because his own efforts — and the disciples that flocked to him — had so advanced the cause of science that he no longer understood a lot of it and was usually out of his depth when the subject was raised. But Edison had never known much about science to begin with — or, more properly, the scientific method. His reading was vast but it was selective, either pursuing a line that had grabbed his interest or serving a project that was already under way. He soon proved himself to be a systematic and effective thinker, but only about the single task at hand, and even then his thinking could not always show him the way. Sometimes his projects were such a mystery to him that he would revert to the child's method of playing with his chemistry set until he found, almost at random, something that worked. In part, he could get away with it because much of science was a mystery even to the scientists of the day. And because he occasionally achieved astonishing results, he came to despise the scientists themselves — the "the-o-ret-i-cal scientists," he called them, Midwestern glottal stops on prominent display. Although he was not an engineer — when he tried to practice engineering in his doomed mining venture, he was so deficient in basic engineering skills that his buildings fell down — the classical contrast between the engineer and the scientist is probably the best way to think of him. A scientist may discover a principle — Newton's first law of motion comes to mind — without suspecting that he has solved an engineering problem. An engineer may solve a problem without suspecting that he has discovered a principle — and on two occasions Edison had been on the verge of a discovery that would have qualified him for a Nobel.

In Newark in the early 1870s, he and Batchelor had observed unusual sparks that passed effortlessly through objects, distance, and flesh, and although he fiddled with them for months and even speculated that they might play some role in wireless transmissions, he decided to call them "etheric force" and never discovered what they were. They were electromagnetic waves, the basis of radio. Later, pondering a light bulb after he'd finally figured out how to invent one, he noticed that it accumulated dirt in a place where no dirt was supposed to be — on the interior of the glass, amid a vacuum. He studied the matter, reported it, and even went so far as to build the predecessor of a vacuum tube. The fact that dirt accumulated in a place where no dirt was supposed to be — in a light bulb — was called the Edison Effect, and he thought it might be of some use as a voltage regulator. In fact, he had observed the action of the electron, a particle that was not discovered until 1897 and then became the basis of all modern electronics. The filament was shedding electrons; this is where the dirt came from. But Edison had no idea what was happening.

On the other hand, he got results. By 1877 he probably had as much hands-on experience with electrical circuits as any scientist on earth, and he had an intuitive sense of how things worked. Unlike most scientists, he also had a sense — partly learned, partly intuitive — of his marketplace, although as the phonograph demonstrated, he could be a little lacking in follow-through. The central power station, if he could pull it off, would liberate him from Western Union and Jay Gould forever. Looking into this new business of the arc light was more than a passing interest.

Fortunately, William Wallace, Moses Farmer's business partner, was nearby in Ansonia, Connecticut, then the brass-manufacturing capital of the country. In Wallace's factory Edison saw eight arc lights of 500 candlepower, which was an enormous amount of candlepower, and inspected a Wallace-Farmer dynamo of 8 horsepower, one of which he ordered. He seemed, as always, his boyishly enthusiastic self. He was also thinking hard.

"I don't think you're on the right track," he said as he left. Something had occurred to him.

* * *

When he returned from Wyoming after observing an eclipse of the sun with some Eastern scientists in the summer of 1878 (and testing a sensitive new heat meter he had invented), Edison was a man reborn. A photograph taken before the trip shows a face lined with the exertions of his phonograph labors, but now he seemed a boy again, swept up in his enthusiasm for the device — the dynamo — that Wallace called a "telemachon." It was the first nonbattery power source he had ever seen. With it, Wallace lit up eight arc lights totaling a staggering 4,000 candlepower. Edison had to have one. His mind was racing. And, as often happened when his mind was racing, he allowed his tongue to race in the presence of a representative of the press. "It was all before me," he said. "I saw the thing had not gone so far but that I had a chance. I saw that what had been done had never been made practically useful. The intense light had not been subdivided so that it could be brought into private houses." In six weeks' time, he declared, he would do just that. "[Other inventors] have all been working in the same groove," he said,

> and when it is known how I have accomplished my object, everybody will wonder why they have never thought of it, it is so simple. When ten lights have been produced by a single electric machine, it has been thought to be a great triumph of scientific skill. With the process I have just discovered, I can produce a thousand — aye, ten thousand — from one machine. Indeed, the number may be said to be infinite. When the brilliancy and cheapness of the lights are made known to the public — which will be in a few weeks, or just as soon as I can thoroughly protect the process — illumination by carbureted hydrogen gas will be discarded.[2]

He meant every word of it. "Carbureted hydrogen gas" was the gaslight of the Sherlock Holmes stories. It was also the gaslight in every major city in America. Some people — the people running the gas companies — were going to be very upset if Edison succeeded.

Few inventions have inspired more myths than the electric light. Edison did not set out to invent a light bulb. Instead, he set out to invent an electric version of the gaslight and, like Brush, he found his inspiration in the experiments of Humphry Davy. Before he invented the arc light, Davy had demonstrated a prototype of the light bulb by making a filament — a thread — of platinum glow. It was yet another demonstration of Ohm's Law. Platinum had a high resistance to electricity. Because of this resistance, some of the electricity converted itself into heat as it flowed through the metal. The heat made the metal glow. It was this experiment, Edison believed, that showed him the way. First, he would devise a simple filament called, at this early point in his thinking, a burner.

Probably it would be made of platinum, Davy's metal of choice, or an alloy of iridium and platinum, another high-resistance substance; a few easy experiments would explain it. It would be exposed to the air, as Davy had done. Edison saw no problem with this, because the electrically heated platinum had a high melting point, and it did not combine with the oxygen in the atmosphere to form a kind of platinum rust that would weaken it. He would place his filament in an existing gas lighting fixture, an S-shaped brass gas pipe that projected from a wall and terminated in a porcelain mantle designed to insulate the metal of the pipe from the heat of the flame. In the mantle, he would place his filament. His electrical wires would be threaded through the pipe. His power source would be an off-the-shelf Wallace dynamo. And his secret weapon, the simple thing that had eluded all the others, would be a thermomagnetic (magnetism triggered by heat) or electromagnetic make-or-break switch, a regulator that would interrupt the current when the temperature of the platinum approached the melting point and would reestablish the current as the temperature fell. In the interval, he calculated, his filament would continue to glow; his electricity would be intermittent, but the light from his burner would be steady because the platinum would be heated to a high temperature that would not fall significantly while the electricity was turned off. With his deep knowledge of telegraph circuits, he believed he could thus contrive matters to achieve a steady illumination. Anyway, people were accustomed to flickering light, they had been accustomed to it for thousands of years, and if his light did flicker a little, people would probably tolerate it.

The subdivision of light — his word for taking the arc light's blinding glare and channeling it into a thousand sources of illumination (his burners, scattered around a house, and the illuminated houses scattered around a city) at gaslight's familiar and user-friendly 8 to 20 candlepower — was thought to be impossible by no less an authority than Scotland's William Thomson. Edison believed it would be child's play. His regulator, the make-or-break switch he would contrive, would do it. He would also confront another form of subdivision. As reflected in Ohm's Law, electricity flowing through a filament expressed itself as light, the desired result, and heat, which is waste. To prevent electricity leakage through heat loss when the customers of an arc light system began to turn off their lamps, the employees of an arc light company would divert the surplus power into spare arc lights in the dynamo rooms while the steam engines powering the dynamos were throttled back. This practice also prevented, although crudely and not well, the lighting magnate's worst nightmare: a customer electrocuted by a sudden power surge. In his enthusiasm, Edison seems to have given no immediate thought to this problem at all, although he should have. As an expert on electricity, he had to know it was an extremely dangerous thing that he was planning to introduce into the nation's homes. It was not hard to imagine a slightly unusual sequence of events — a wet floor, bare feet, and a finger in a light socket — that would render a customer dead before she hit the ground, even if the power wasn't surging. But his vision had gotten the better of him, he entirely ignored electricity's darker side, and he proposed to be up and running in a double handful of days. He told his friends in the press that he could illuminate all of lower Manhattan with one of Wallace's dynamos and a 500-horsepower steam engine. He was certain of it.

When the newspapers hit the bricks bearing Edison's confident, happy words, the price of gas shares collapsed on the London exchange.

The Americans remembered Edison in his previous role as the sidekick of Jay Gould, the least trustworthy man in the country, but when they thought of him as an inventor, they regarded him with a kind of superstitious bafflement. The British, by contrast, held him in little short of awe. He seemed to be American genius personified, a latter-day Franklin, a self-taught Brother Jonathan who worked miracles. In reaction to his announcement, the British formed a parliamentary committee to examine his latest claims. Edison the unstoppable had become the subject of statecraft, but for once, the parliamentarians concluded, he appeared to be barking up the wrong tree. In other words, they thought he was wrong. Edison's project, the committee concluded, was "good enough for our transatlantic friends," but it was "unworthy of the attention of practical or scientific men."[3] The electric light, after all, had been around for seventy-six years, many people had been fiddling with it, and the arc light, first demonstrated by Davy in 1809, was its only fruit. Davy had also demonstrated the possibilities of platinum, the French had even built a platinum-filament vacuum bulb in the 1820s, and in 1840 an entire English theater had been dimly illuminated with them — at an unfortunate cost of hundreds of pounds for every kilowatt-hour. Moses Farmer, the Connecticut inventor, astounded the world in 1859 by lighting up his home with just such platinum-filament bulbs — the same bulbs Edison had seen in the Boston shop window — but nothing had ever come of them. Farmer had also been fooling around with carbon-filament bulbs and for years so had England's own Joseph Swan, a chemist who also went on to invent photographic paper. Platinum was hideously expensive, carbon soon burned up, and nothing lasted very long. Now Edison, with his air-exposed platinum burners, was going all the way back to Davy. If Edison's device worked, Sir William Thomson told the parliamentarians, it would be so dim as to be useless. The legislators concluded that there was nothing new here, and a long record of failure. England had nothing to fear from Edison's new venture.

Wall Street wasn't so sure. On September 17, as Edison was gearing up for what he confidently expected was a quick-and-dirty spasm of successful invention, he received a telegram from his lawyer, Grosvenor P. Lowrey. Lowrey regarded the brilliant, self-centered, and eccentric Edison as a kind of scientific pet as well as a lucrative client. He was also extremely well connected with the New York money interests, and the New York money interests had read the same newspapers as the London politicians. Unlike the politicians, the American moneymen didn't have the advantage of Sir William's wise counsel — they didn't know much of anything useful about electricity, science, or the history of the electric light — but they had seen Edison come out of nowhere before, out of nowhere and in the wrong company.

In the autumn of 1878, Grosvenor Lowrey's task was twofold: to manage the publicity about Edison's new platinum burner, and then to sign up the investors thus attracted, whoever they might be. The first part more or less took care of itself. Edison was almost always eager to talk to the press, and the press regarded him as excellent copy. Repeatedly he crowed about the new electric light he would produce the day after tomorrow if not sooner. Repeatedly he announced that it would be a "big bonanza." In this phase of the operation, Lowrey's job was to get out of Edison's way and let him talk. Meanwhile he began to collect some of the investors. Western Union, mindful of how it could once have bought Bell's telephone for a pittance, invested $30,000. William H. Vanderbilt, his son-in-law Hamilton Twombly, the J. P. Morgan partner Eggisto Fabbri, and various Western Union executives together put up another $50,000; the Vanderbilt fortune was heavily invested in gas utilities, and it was hedging its bet. Other investors kicked in an additional $70,000. And Morgan himself came in, which meant the rules of the game had suddenly become very clear.

J. P. Morgan was a financier dedicated to orderly markets and a higher ethical standard than many of his contemporaries, and he had dedicated his life to defeating Jay Gould and all his works. With Morgan lending the lighting venture his name and vast resources, Edison would not be permitted to fall into Gould's hands again. No, it would be time to place him — cautiously, for Morgan was not a careless man — in the right hands. Perhaps something would come of the platinum burner. If nothing came of it, the Morgan interests would win the mercurial inventor's gratitude — and Edison was the sort of man who kept coming up with amazing things. There was certain to be money in at least some of them, and Morgan, Edison's benefactor, had prudently placed himself on the ground floor.

Here is the origin of the second myth about Edison and the electric light: that he became the creature of the formidable J. P.

Morgan. (The first myth is that Edison set out to invent the light bulb. He didn't.) Morgan was not a robber baron; he was the proto-type of the modern investment banker. He was the sworn enemy of the likes of Gould and the friend of gentlemen much like himself. He was ferocious, mercilessly logical, and — in modern parlance — a gadget freak. Edison's electricity called to him. He was deter-mined to make his new home on Madison Avenue the first New York dwelling equipped with the new lights. And it was. More to the point, Morgan was the dominant financier of his time, the only financier with a higher education, and one of the few financiers who was born rich and socially prominent. When Morgan came forward with money at a difficult time in the electric light's fortune, the other Edison investors, among them the Vanderbilts, steadied. When Morgan and his partners made personal investments in the light, the other investors took heart. And unlike Gould, Morgan did not exact tribute. His firm, Drexel Morgan, received the exclu-sive right to sell the new light in Australia, New Zealand, and Por-tugal, not exactly hot properties in 1880. They would sell the light at their own expense. Edison would receive more than half of all the proceeds. In all, the Morgan firm and its partners invested no more than a million dollars, a generous but neither an excessive nor a controlling sum, and in Edison's opinion, Drexel Morgan's loans were awarded with a somewhat stinting hand. Morgan became Edi-son's personal banker, never his master, as history may have led us to believe.

Edison was assigned half the shares in the new Edison Electric Light Company, and the new enterprise was capitalized at $300,000. Even factoring in 120-odd years of inflation, it was not a lot of money, especially in view of the resources of the investors; the modern equivalent of the company's capitalization would be three million dollars. And here, although none of them knew it at the time, was sown the seed of a very large cross-purpose. The investors looked forward to rich royalties when Edison completed his inven-tion and licensed it to manufacturers. And Edison looked forward to complete, painless, and very swift success, but in a very different form. First, he would invent the light. Then he would devise the electrical power system to illuminate it. He would manufacture the bulbs. He would build the system. The idea of sitting back and collecting royalties played no part in his thinking at all.

At Menlo Park, he buckled down to work. "The ordinary rules of industry seem to be reversed at Menlo Park," the *New York Herald* reported on January 17, 1879.

> Edison and his numerous assistants turn night into day and day into night. At six o'clock in the evening the machinists and electricians assemble in the laboratory. Edison is already present, attired in a suit of blue flannel, with hair uncombed and straggling over his eyes, a silk handkerchief around his neck, his hands and face somewhat begrimed and his whole air that of a man with a purpose and indifferent to everything save that purpose. By a quarter past six the quiet laboratory has become transformed into a hive of industry. The hum of machinery drowns all other sounds and each man is at his particular post. Some are drawing out curiously shaped wire so delicate it would seem an unwary touch would demolish them. Others are vigorously filing on queer looking pieces of brass; others are adjusting little globular shaped contrivances before them. Every man seems to be engaged at something different from that occupying the attention of his fellow workman. Edison himself flits about, first to one bench, then to another, examining here, instructing there, at another earnestly watching the progress of some experiment. Sometimes he hastily leaves the busy throng of workmen and for an hour or more is seen by no one. Where he is the general body of assistants do not know or ask, but his few principal men are aware that in a quiet corner upstairs in the old workshop, with a single light to dispel the darkness around, sits the inventor, with pencil and paper, drawing, figuring, pondering. In these mo-

ments he is rarely disturbed. If any important question of construction arises on which his advice is necessary the workmen wait. Sometimes they wait for hours in idleness, but at the library such idleness is considered far more profitable than any interference with the inventor while he is in the throes of invention.[4]

It has often been said that Menlo Park was the prototype of the institutional research lab. At a research lab, independent minds pursue independent lines of work. At Menlo Park, by contrast, the Tabernacle was organized along the familiar lines of the machine shop, where Edison's top cronies like Batchelor and Kruesi were also his employees. When Edison was absent or stopped, work ground to a halt. There was only one independent mind at Menlo Park.

Nor was Menlo Park soon to be unique in either its funding or its resources. In New Britain, Connecticut, where an arc light business would be funded, rather oddly, by the local locksmiths, Edison would be closely watched by Elihu Thomson, the latest in the line of Fighting Thomsons, none of them apparently related, to fool around with electricity. Elihu Thomson, another great American scientist of whom no one has ever heard, owes his obscurity to a singular fact: although he once strangled a cat, he was one of the nicest guys who ever lived. The title of his biography says it all: *Elihu Thomson, Beloved Scientist*. Nice guys seldom make good newspaper copy, and they seldom capture the attention of posterity. Thomson did not cultivate Edison's gift for self-promotion, and he was not interested in fame, but in electrical inventiveness Thomson was Edison's equal if not superior. Edison wrote his patents sloppily and patented everything in sight, even if it didn't work. Thomson's patents were models of the form. Unlike Edison, he never took out a patent for the hell of it; Thomson's patents were for useful devices that were carefully thought out and invariably worked. Moreover, Thomson soon knew far more about electricity than Edison did, which placed him in an almost unique category. In 1878 he was still a Philadelphia high school teacher, but by the end of the 1880s he possessed far greater resources than Edison, thanks to the shoe manufacturers of Lynn, Massachusetts, who bought his company, moved it, and placed one of their number in charge, allowing Thomson to stick to his inventive last.

Thomson had already invented something Edison badly needed: a self-regulating dynamo, a dynamo that automatically adjusted itself to the line load. It regulated the amount of electricity in the wire and did not hunt, or go out of synchronization with other dynamos. His New England company, Thomson-Houston, was one of the largest electrical manufacturers in the country, his arc lights were the finest in America, he had a sophisticated knowledge of currents, and he knew a great deal about systems management. Thomson had also invented a type of transformer that distributed the power, that is, chose which wire to send how much electricity down, something Edison never did. Some form of relationship, perhaps a business alliance or perhaps an outright purchase of Thomson's technology, was clearly indicated, but Edison, plunging forward, gave no indication that he even knew Elihu Thomson existed.

With a high heart Edison advanced upon his quarry, the electric gaslight. The task ahead still seemed simple. He believed he had only to discover the proper iteration of his proposed make-or-break regulator, and platinum remained his metal of choice for his burner, although he also fell back into his preferred and seemingly random try-anything approach. He tried a platinum-iridium alloy, ruthenium, aluminum, silicon, chromium, molybdenum, palladium, boron, titanium, and manganese. Moses Farmer helpfully sent along a bar of iridium and — prophetically — suggested carbon in a vacuum. Edison had previously experimented with carbon. He believed it would burn up after only a few minutes, and Farmer's suggestion was ignored. For a while, nickel seemed even more promising even than platinum, and it had the advantage of being cheap. Edison crowed about a breakthrough, and Lowrey told him for God's sake to shut up; the investors were growing weary of Edison's endless announcements of success just as soon as a few trifling obstacles were overcome. Lowrey thought that a shift into nickel after all the bold talk about platinum would alarm them further. It would also look as though Edison were thrashing around, which he was. And Lowrey was correct: nickel proved to be another dead end.

Dead ends, in fact, were all Edison had to show, and toward the end of the year the experiments at Menlo took a dramatic new turn. An instrument not much in use at the lab, the microscope, was trundled out, and Edison began to practice science. He decided he had to understand what was happening to the filaments on a molecular level. The filaments would glow and burn, and then they would burn out. The filaments, he found with the microscope, were cracking and bubbling and outgassing. Hydrogen seemed to be the culprit. The problem was no longer — exclusively, that is — the melting point of the metal. The solution, Edison concluded, was a glass bulb and a vacuum. This was not a novel idea. Countless other experimenters had hit upon the same idea, including Swan and Farmer, but none of them had ever been able to make it work. The problem was creating the vacuum. Nobody had ever made a vacuum that was good enough to prevent a filament from burning itself up.

Meanwhile, Edison was busy elsewhere. He had always insisted that an electric light would be cheaper than gas and an open flame but he had no evidence of any sort that the statement was true. Another of his preferred research methods was to read everything relevant that he could lay hands on. He subscribed to dozens of gas journals and trade magazines, pored over them, and filled sheet after sheet of paper with computations that seemed to prove what he already believed: that electricity would in fact cost less than gas. But until he settled on his burner and chose or invented a dynamo to power it, it was a pointless exercise. He had no idea whatsoever how much this electrical system would cost. Then there was the problem of the copper wire that would convey the current from the dynamo to the customer. Edison was planning to use direct current at low resistance, everybody's preferred method of conveying electricity except for a handful of Europeans, including the Russian Paul Jablochkov, who had lit up the arc lights in Paris, and Elihu

Thomson. But low-resistance direct current distributed over a wide area meant that the wire leading from the dynamo would need an immense diameter that would gradually taper the further it went. Direct current flowed in a steady stream. The further away your customers were, the more electricity you had to pump into the line, and the only way to pump in more electricity was to increase the diameter of the wire. Copper was expensive. And platinum, still Edison's preferred metal for the burner, was even more expensive. When properly considered, the notion that Edison's electrical system would be cheaper than gas was laughable.

He had always spoken of creating a system. It was a logical extension of the whole idea of creating electrical illumination. Edison couldn't sell his burners — later, his bulbs — unless the customers had access to electricity, and in the Edison mind, he was the natural person to supply it. The electricity would be generated in a power station containing a number of dynamos. It would then be distributed to the customers, either in wires on poles or in wires buried in the ground. The electricity generated by the dynamos would be regulated to meet the needs of the customers. In some way, the electricity consumed by the customer would have to be measured, so the customer could be billed. The measuring device must be read on a regular and scheduled basis. Clerks must compile the bills. The bills had to be sent. An entire system was needed. Here Edison parted company from the run of electrical inventors, like Swan, who had concentrated on the electric light and used a co-located source of power — an individual battery hooked up to each individual lamp. Edison had never met Swan, but he soon came to know of his work. Wallace's dynamo had suggested a system running off a central power station, and Wallace had always mentioned a system when he mentioned his light, although it is hard to know if he mentioned it to Edison, either when they met or in their correspondence. The idea, vague though it was, was not, as is commonly believed by his biographers, unique to Edison. Charles Brush had already built such a system, complete with a central power station, in San Francisco.

Nor was the idea of distributing electricity over a wide area another of Edison's unique inspirations. Brush had already done that, too, in 1879 in Wabash, Indiana, where he lit up the entire town with arc lights. Edison apparently never met Brush, although Brush was building arc lights in Manhattan. And Edison didn't know much about dynamos. Until now, he'd had no reason to.

Inventing a new kind of dynamo, however, proved to be a snap, especially when compared to inventing an electric light. To Wallace's consternation, Edison decided that the Wallace dynamo was not equal to the task of powering the system he envisioned. It could illuminate only three lights, and Edison wanted to double that. He ordered a Gramme dynamo and then a Siemens one, took them apart, and began to experiment with armatures and windings. In a dynamo, the shape of the armatures, their position in the magnetic field, and the way the copper wire was wound determined the generation of the electricity, how much you got, and with what efficiency. He had little to go on, and nobody else did, either. But Edison had always had a deep instinctive understanding of electrical machinery, was enormously self-confident, had the diligence of the bee, and had a valuable new assistant. A university-trained physicist, young Francis Upton was originally hired to research the literature and the patent situation to give Edison a better idea of the electrical art and its state, and Upton could do his work by using math, which saved a certain amount of time. Edison nicknamed him Culture. Together they built the dynamo dubbed the Long-Legged Mary Ann because its two vertical electromagnets made it resemble a girl with her legs in the air. The efficiency of a dynamo — the amount of steam power converted into electricity — was then around thirty-five percent. Observing the situation from afar, Elihu Thomson noted with admiration that the efficiency of the Long-Legged Mary Ann was ninety percent. Batchelor called it a "Faradic machine," after Faraday, who had discovered how to generate electricity.[5]

So they had a dynamo, and it was a very good one indeed. But they were beginning to run out of money. Edison had spent the first

$50,000 of his investors' money to build a new brick machine shop behind the Tabernacle, and it was going to cost $18,000 to wire Menlo Park itself. No fresh funds would be forthcoming from the existing investors. Edison's six weeks to invention had long since expired. He had burned his way through the final months of 1878, and although his capitalists had not withdrawn, they were not in a generous and open-handed mood. Lowrey turned to Morgan. Morgan produced the necessary funds. Then he demanded a demonstration at Menlo Park to see what he had bought. The platinum filaments glowed faintly in their new glass bulbs, just as William Thomson had predicted to Parliament. Although they were new, several of them burned out. Wordlessly, Batchelor replaced them. Wordlessly, Morgan left. And he let his money ride. The prototypical investment banker had just invented another new profession for the twentieth century. He was now the prototypical venture capitalist.

After the turn of the year 1879, Edison had two principal concerns: creating a vacuum and finding the proper length for his platinum filament. He continued to search for an insulating medium — he tried manganese oxide — that would prevent the platinum from cracking and melting, and he incorporated various forms of regulator in his new bulbs, including one so fiendishly complex that it's impossible to imagine how it could have been manufactured. As usual, he patented everything in sight. There seemed to be no intrinsic reason for platinum to function better in a vacuum — oxidation was never a problem — but it did. Edison bought one of the new Sprengel vacuum pumps, a complex of mercury-filled glass tubes that was taller than a man, and combined it with the older Geissler pump and a master glassblower. Batchelor continued to experiment with the shape of the filaments. He had wound them around a core of chalk. He wound them around bobbins. And, with Edison's second great breakthrough after the vacuum tube, he found himself shaping ever-longer lengths of wire into increasingly complex coils, trying to find the proper shape for the filament.

Like everyone else who experimented with electricity, Edison originally believed that the resistance in the circuit should be equal to the resistance in the dynamo; it was the way things had always been done. Then something occurred to him: if the resistance in the dynamo remained low but the resistance in the part of the circuit represented by the filament was dramatically raised, the system became dramatically more efficient. He tried it out; Upton, who thought he understood Ohm's Law, was astounded by the result. Dramatic efficiency meant a brighter light. It also meant that the amount of copper in the system fell by half. Edison, though he was not the first, had discovered the miracle of voltage. By increasing the velocity of electricity, he could run more of it down his wires. But a higher resistance in the circuit meant that the filament had to be changed in both length and diameter. Edison calculated that a sixteen-inch piece of .009-inch platinum wire at 62 degrees Fahrenheit had a resistance of two ohms. If the diameter was narrowed to .001125, the resistance would rise to 640 ohms, and if the length was doubled, the resistance would hit 1280 ohms, which, indeed, improved the functioning of the circuit. Following Edison's lead, Batchelor spent most of the rest of the year trying to make spirals of extremely long, thin wire. And Edison pursued what he believed to be the last step before the bulb was perfected: finding a suitable insulator. Both activities were a complete waste of time. But by October, the improved Sprengel was able to produce a vacuum of a millionth of an atmosphere. In his small, fragile bulbs, Edison was able to create the most perfect vacuum in the world. But weeks — months — passed, and there was no further progress.

In the workroom they sang, after Gilbert and Sullivan's hot new 1878 musical HMS *Pinafore*:

> I am the wizard of the electric light
> And a wide-awake wizard, too,
> Quadruplex, telegraph or funny phonograph
> It's all the same to me.
> With ideas I evolve and problems that I solve
> I'm never, never stumped, you see. [6]

On the zither, Edison's new glassblower played the lugubrious bar-room ballads Edison so dearly loved: "My Heart Is Sad with Dreaming" and "The Heart Bowed Down."

* * *

Concerning the revolutionary events that transpired in October 1879, Edison always told a story that conformed to the public's romantic notion of the inventive process rather than the truth, and it is this story — that Edison conceived the light bulb in a eureka moment, that he never conceived of anything but a light bulb, and that it was invented on a specific day — that has found its way into the nation's folklore and history books. The reality was very different.

By October, Edison had built his radically improved dynamo. He had begun to mentally form the outline of his system, based on a central power station, although when he turned his thoughts to New York City, where he proposed to build it, his notion of political reality was somewhat naïve. He had created a usable vacuum, the one thing that had prevented Joseph Swan from beating Edison to the light, not by months but by more than ten years. He had discovered the signal importance of creating high resistance in his circuit, which required the filament to be wound in spirals. Resistance in the circuit made the filament glow in a highly satisfactory manner, but otherwise, no progress had been made. The all-important insulator for the platinum filaments remained elusive. It was, he thought, the last obstacle.

It's unclear how carbon entered the picture. Edison claimed that he'd experimented unsatisfactorily with the substance as long ago as 1877 and given it up as a bad job. Moreover, his extensive reading on the subject revealed that a carbon filament had occurred to just about everybody who fooled around with the possibility of electric light, including Humphry Davy, and absolutely none of them could make it work. He told the *New York Herald* reporter Marshall Fox that the sudden return to carbon was an accident. "Sitting one night in his laboratory reflecting on some of the unfinished details," Fox wrote, "Edison began abstractedly rolling between his fingers a

piece of compressed lampblack until it had become a slender filament. Happening to glance at it the idea occurred to him that it
might give good results as a burner if made incandescent."[7]

It's as good a story as any and, in fact, it's the only story we have.
Edison was not always a truthful man. "Made a mould for squeez-
ing," Batchelor wrote in his lab notes for October 7,

> put in some Wallace soft carbon, and squeezed it out of a
> hole .02 [inch] diameter getting it out a yard long if re-
> quired. Could make even more sticks by rolling on glass
> plate with piece of very smooth wood. These sticks could
> be rolled down to .01 and then wound in spirals. We made
> some and baked them at a red heat for 15 minutes in a
> closed tube. When taken out they were hard and solid
> much more so than we expected and not at all altered in
> shape. A spiral made of 'burnt lampblack' mixed with a
> little tar was even better than the Wallace mixture. With
> a spiral having 5 inches of wire of .01 we can get 100
> ohms. We now made a double spiral on brass so as to wind
> the carbon so similar to some of the first platinum spirals
> we made.[8]

There is no clue as to the fate of these first carbon filaments. In any
event, there was very little excitement over the new substance and
very little action. On October 11, Upton measured the resistance of
a stick of carbon with a Wheatstone bridge, a metering device
named for Charles Wheatstone but actually invented by a member
of the Christie art-auction family, and got disappointing results.
Over the next few days, the lab tried to insulate platinum with sil-
ica and dealt with trouble at the vacuum pump. Then everybody
but Upton and another new hire, Francis Jehl, went off to deal with
a problem involving the chalk drums destined for Edison's telephone operation in England.

None of them realized how close they were. "A stick of carbon
brought up in a vacuum to 40 candles (say)," Upton wrote in his lab

notes on October 19. "Mr. B. trying to make a spiral of carbon. Grease in the pump. Gauge broken. Trying to make carbon spirals." On the twenty-first, Batchelor returned to the problem, making more spirals out of tar and lampblack putty and carbonizing them in an oven. Under the heading "Electric Light Carbon Wire," Batchelor wrote:

> A spiral wound round a paper cone no matter how thin always breaks, because it contracts so much. If the heating is done slowly this is modified but with the present proportion of tar and lampblack it will always break. One of the great difficulties is to keep the spiral in position whilst you carbonize it. This might be remedied to a great extent by using a hollow sleeve winding the spiral inside with something to hold the ends whilst they are being fastened to the leading wires.[9]

October 21, the traditional day of the electric light's invention, passed without incident. On the morning of the twenty-second, Batchelor wrote: "We made some very interesting experiments on straight carbons made from cotton thread so." Upton wrote:

> Carbon spirals and threads. Trying to make a lamp of a carbonized thread. 100 ohms can be made from an inch of .010 inch thread. A thread with 45 ohms resistance when cold was brought up in a high vacuum to 4 candles for two or three hours and then the resistance seemed to concentrate in one spot. Resistance cold 800 ohms.[10]

They were almost there. Batchelor made a list of the filaments they tried:
  1 — of vulcanized fiber
  2 — Thread rubbed with tarred lampblack
  3 — Soft paper
  4 — Fish line

5 — Fine thread plaited together 6 strands

6 — Soft paper saturated with tar

7 — Tar & Lampblack with half its bulk of finely divided lime work[ed] down to .020 — straight one 1/2 inch

8 — 200 is 6 cord 8 strand

9 — 20s coats 6 cord — no coating of any kind

10 — Cardboard

11 — Cotton soaked in tar (boiling) & put on[11]

If there was a moment when the electric light was born, it came when number 9, a carbonized piece of standard #6 Coates thread burned for thirteen and a half hours. There was no celebration. Everyone kept right on working. A deeply fragile piece of carbonized sewing thread was an absolutely terrible filament. Furthermore, it wasn't a spiral, Edison's preferred configuration. Over the next week, Batchelor tried celluloid, boxwood shavings, spruce, hickory, baywood, cedar, rosewood, maple, punk, cork, flax, coconut hair and shell, and paper. It wasn't until the middle of November that they settled on the filament they would use for the public demonstration Edison was planning: a simple horseshoe of carbonized cardboard. It would not, of course, be the final filament, but it would do in a pinch.

"The Electric Light is slowly advancing from the last big step," Upton wrote to his parents.

> We now know we have something and that is what we [did] not know until last week. We can compete with gas in a great many ways now though not as completely as we wish, yet there seems nothing to prevent our getting a perfect burner that shall do as well as gas. Time and cost will prove what we have to be good or bad.

Hardly inventive triumphalism. They still had a job of work to do.

Edison, who never met a reporter he didn't like, now almost totally shut down on the press. Perhaps he felt he'd cried wolf too often or perhaps he was relishing his surprise. To his representative in

England, Colonel George Gouraud, he cabled: "Say nothing." He asked Western Union to send over two linemen; Menlo Park's main drag, Crystie Street, was to be lined with the electric lights. He ordered steam engines for his dynamos. Upton's house was wired in early December. Edison's house soon followed. The only reporter Edison was talking to was Marshall Fox. On December 18 or 19, it was announced that a public exhibition of the new light would be held on New Year's Eve. Fox broke his story on the twenty-first. "Mr. Edison," Upton wrote his parents, "is very much provoked and is working off his surplus energy today." Edison also cabled London to get a move on with the European patents.

The rejoicing was not exactly universal. A number of editors remembered that Edison was the Professor of Duplicity, and they so reminded their readers. "Three times within the short space of eighteen months," said the *Saturday Review* of London,

> [Edison] has had the glory of finally and triumphantly solving a problem of worldwide interest. It is true that each time the problem has been the same [but] there is no reason why he should not for the next twenty years completely solve the problem of the electric light twice a year without in any way interfering with the interest or novelty. There is a strong flavor of humbug about the whole matter. Mr. Edison's efforts in electric lighting seem cursed with a total absence of originality. He is an inventor who is absolutely intoxicated with his own reputation.

And, others pointed out, he had invented the phonograph and done nothing with it.

He had indeed done nothing with the phonograph, but he was about to make the *Saturday Review* eat its words. Morgan and the board voted to give Edison $75 so he could buy a suit, and at last the great day was at hand: New Year's Eve 1879. The Pennsylvania Rail Road again piled on extra trains; three hundred people flocked to Menlo Park to see the miracle with their own eyes. There were two globes at the door of Edison's office, eight on poles outside, and thirty in the lab. Edison also lit up a bulb underwater, and it burned for four hours. Several men who wandered into the dynamo room found their pocket watches magnetized. In the course of the event, eight of the bulbs were stolen. A gas company operative arrived with a jumper cable threaded through the sleeves of his coat. When he tried to short-circuit the system, only four lights went out. Without telling anyone, Edison had invented the fuse.

* * *

Yes, it was all quite miraculous, but now he had to do something with it. He had always planned to install his system in New York. New York money had backed him. The New York papers were his preferred outlet. Edison was a household name, thanks to New York. But to build a system in New York, he had to do three things. First, he had to obtain the permission of the city fathers, which meant an appeal to Tammany Hall, the corrupt political organization that ran the city. Second, he had to figure out what a system was. He was well on his way here, but a system carrying nothing but electricity would probably have some unique demands that had not occurred in a gas system or a telegraph system. And finally, he had to do it on his own nickel. His investors — Morgan being the one exception — expected to sit back and reap rich rewards from licensing fees. Building an electrical system, and the factories to support it, was nowhere in their plans.

Edison had always planned to model his system on the gas mains. Gas mains had always been buried. Although lower Manhattan was festooned with wires on poles — wires for the gold machines, wires for the stock tickers, and now wires for the telephones — and digging in the city was ruinously expensive, Edison planned to bury his wires. At least part of his reason was to avoid bad publicity: every so often a lineman was electrocuted in full public view, slowly cooking to death as the spectators looked on, and Edison was determined to avoid that. In the public mind, electricity was a dicey enough proposition without dead linemen, and Edison planned to emphasize the wonderful safety of his new product, reminding his future customers that a light bulb, unlike a gas flame, could not be blown out. "In those days," wrote his future secretary, Alfred Tate, "in every hotel bedroom a printed card hung from the gas fixture, inscribed with the legend 'Don't blow out the gas,' notwithstanding which the gas was not infrequently blown out with the result that the vital flame of the occupant of the bed also went out a few hours later." Edison the blowhard inventor was about to be replaced by Edison the showman who had made a small circus out of the introduction of the phonograph. The methodical Upton made a task list of the computations that lay ahead.

How many feet of gas are consumed in various parts of N.Y. per consumer and an estimate of the average time of burning.

How much horsepower is taken and the price paid per horse-power up to 10 horsepower.

Cost of pipes laid in the street and depreciation.

Cost of management.

Cost of building lots or rent of same.

Cost of water for H.P.

Houses — estimated cost of introducing.

Cost of condensing as compared to high pressure engines.

Cost of laying pipes or wires in the street.

There was no want of eager hands to take up the work. Ever since Edison rolled out the electric light, eager young men had been flocking to Menlo Park to work for a song. There was other work to be done. The Long-Legged Mary Anns were improved with new commutators. A new configuration was set for the brushes that picked up the electricity and conveyed it to the wires. Attempts were made to bleed more of the wasteful eddy currents out of the armature. Edison was not satisfied with the light bulb, particularly the cardboard filament. It wasn't consistent. One bulb had burned for 550 hours, another burned for close to 500 hours, and still others blew out the moment they were plugged in. Cardboard simply wasn't a homogeneous substance; at the molecular level, too much was going on, and Edison assigned the search for a new filament to his chemist, John Lawson, whom Edison the compulsive nicknamer called Basic. Lawson tested flour paste, leather, macaroni, sassafras, pith, cinnamon bark, eucalyptus, turnip, and ginger root. It was searches like this that gave Edison a bad name as a random tinkerer, but for the most part Lawson's search was methodical and disciplined. Edison wanted some form of vegetable mass with straight and parallel fibers, and it was there that Lawson concentrated the majority of his attention. Months later, Japanese bamboo was settled upon, and bamboo remained the filament for years to come. Still, Edison the showman couldn't rest. He dispatched agents — including a New Jersey schoolteacher who circumnavigated the globe — to scour the earth for filament material, and he made certain that the papers carried full accounts of their sometimes hair-raising adventures in the wild parts of the planet. Later he formed some of his men into a phalanx and marched them down Broadway with light bulbs atop their derbies, the power coming from a portable dynamo in a cart concealed in their midst; as their leader they had a drum major on a tall charger, twirling a light-bulb-tipped baton. Then there was the Edison Darkie, a comical black man. The Edison Darkie, too, had a light bulb atop his head. It lit up when special pointed contacts in his shoes touched wires concealed in the rug as he handed out pamphlets extolling his master's creation. As previously noted, Edison was extremely fond of what were called coon jokes, and the Edison Darkie was one of them.

He was not to be rushed. The new filament had yet to be found, and he was trying to purge the last remaining platinum from his lamps: the little, and extremely expensive, catches that held the bulb to its base. Platinum was used because of its high melting point. Batchelor was set to finding a carbon substitute. The search failed. But no matter. Everyone, it seemed, wanted Edison's imperfect light bulb, from the high-ranking scientists who flocked to Menlo Park to the great Morgan himself, eager to add electric light to the decorations of his new mansion and thirsting for the world's first desk lamp, a thing Edison had promised would be his. Orders flooded in. William Vanderbilt wanted a steam engine and dynamo in his basement, preferably before Morgan. Edison turned him down. Only to Henry Villard was his ear not deaf.

Like Vanderbilt, Villard was an Edison investor and board member and, in his capacity as the owner of the Oregon Railway and Navigation Company, he was also the owner of the new ship *Columbia*. Fire from open flames was an ancient hazard of the sea. Though the plight of seamen did not normally call to the inventor, Edison the showman surfaced again. The improvements to Long-Legged Mary Ann were finished, the cardboard filament lamps would do, and *Columbia* would become a floating advertisement for his wares. Four Mary Anns were installed, the wiring was crudely insulated with cotton, and each lamp was slowly and painstakingly assembled; Edison had yet to establish an assembly line. *Columbia*, the first Edison installation outside Menlo Park, advertised the wizard's wares for a dozen years to come.

At Menlo Park, not all the work centered on the light bulb and the delivery system. There was an attempt to adapt an electric motor to the chalk button telephone Edison was selling in England, and he began to make drawings of a magnetic ore separator, a device that used electromagnets to extract iron ore from beach sand and the first step toward the disastrous mining venture that would consume most of his time during the nineties. But the light bulb and delivery system were the principal fields of endeavor. An old factory across the railroad tracks at Menlo Park was converted into a plant for making bulbs. A competition erupted among the muckers to devise an improved vacuum pump, 476 of which would be installed. Young Francis Jehl, who had come to Edison through the good offices of Grosvenor Lowrey, devised a pump that would evacuate a bulb in six minutes. Batchelor found a way to straighten the new bamboo filaments, which tended to bend during carbonization. And an attempt was made to eliminate the glassblowers by contriving a bulb mold. The effort failed; the glassblowers remained expensively on the payroll until the work was contracted out to Corning, a company that specialized in making glass. His mind

never at rest, Edison began to turn his thoughts to the next generation of dynamos. They would be large — they would become the famous Jumbos, so named because the first one was shipped to Paris in
the box that had brought Barnum's celebrated elephant to America — and they would be linked directly to powerful steam engines
rather than powered by the traditional belt-and-pulley system that
Edison believed wasted energy. But he was never methodical unless
he had to be. For example, he never attempted to measure the energy he believed was wasted. Despite the statements he made in his
authorized biography — a book supposedly written by a pair of professional authors, Frank Dyer and Thomas Commerford Martin, but
actually written by Edison's own secretary, William H. Meadowcroft — he never worked out the inefficiency problems that arose as
a result of the direct connection. Indeed, they would never be resolved until the steam turbine, a machine that was a steam engine
and a turbine in one, was invented in 1884, not by Edison. Quietly,
despite the huge publicity surrounding the Jumbo, most of Edison's
power stations were equipped with versions of the reliable Mary
Ann, a genuine breakthrough device. And on May 1, ditchdiggers
began to lay out a full-scale model of the power system Edison proposed to build in New York.

Another amazing example of careless and sloppy work: As Edison laid his wires for the exhibition he planned in Menlo Park and
covered them, he didn't bother to test them. Nor, for the longest
time, did he bother to insulate them; he was using a low voltage,
and he didn't believe insulation was necessary. As a result, he electrified everything in sight, including Menlo Park's trees. The whole
setup had to be dug up and done all over again — four times. The
problem, repeatedly, was the miserable insulation of the time. Edison's muckers tried cloth, rubber, wax, anything that promised to
work, separately or in combination, before settling on a malodorous
witches' brew of asphalt, paraffin, linseed oil, and beeswax that was
cooked up in 260-gallon outdoor vats. Downwind, Menlo Park
could be smelled for miles.

As winter approached, a new urgency entered their work. Edison and his team were losing their lead. Charles Brush, the unstoppable, lit up Broadway, which soon became the Great White Way, with his arc lights. The later inventor of the machine gun, Hiram Maxim, installed incandescent bulbs that were clear Edison rip-offs in the vaults of the Mercantile Safe Deposit Company. There were renewed mutterings about Edison in the press. The word "humbug" reappeared. It wasn't until Monday, November 1, that his men were finally able to turn on the street lights in Menlo, and the following night "the entire line along Turnpike from Carmans to Factory was supplied with lamps and burned till nearly 12 o'clock." Edison put out word that the illumination was in honor of the election of James Garfield to the presidency. It wasn't. Meanwhile, he was no closer to illuminating New York. He had his bulbs and his generators, but he didn't have permission to deploy them in his intended market. Understandably, the investors were growing restless. A year had passed since the first successful bulb glowed for thirteen and a half hours, and until Edison got himself into New York, the pockets of the Edison investors' buckskins were going to see no shiny new doubloons. Lowrey explained how they would obtain their New York permit: they would apply as a gas company under the Gas Company Act. It was the only legal way they could get permission to dig up the streets. Then, as the statute permitted, they would apply for permission to lay electrical wiring. There was the rub. Applying for permission meant approaching Tammany, and Edison *still* hadn't done so. And in the New York of 1880, not a sparrow fell in New York City without Tammany's permission.

On Monday evening, December 20, the trick was finally done; by formal invitation, the city's aldermen were welcomed to Menlo Park. Edison readied his system, laid on food, and bought a copious quantity of liquor to wet the Tammany whistle. At the end of the day, a happy alderman declared that "Professor" Edison handled his cigar just like a regular fellow.

The regular fellow was allowed to dig up the streets of New York. To his horror, "Professor" Edison discovered that New York had street inspectors. Moreover, he was expected to pay them a daily fee of $25. Edison had a very bad moment until he discovered that the fee was a payoff. As long as the $125 was paid over every Friday, the street inspectors remained invisible.

* * *

Edison became, as he had once foreseen in an 1870 letter to his parents, a bloated Eastern manufacturer, seemingly forgetting that he had moved to Menlo Park to stop being just that and instead spend all his time inventing something. But the electric light was the end of the Tabernacle. It lingered on for a while, briefly functioning as an actual industrial research lab before falling into a ruin eventually rescued, complete with the dirt it rested on, by Henry Ford. Henry Ford, who was once chief engineer of Detroit Edison, worshiped Thomas Edison, and Edison was content to let him do it.

Edison took new offices in a brownstone at 65 Fifth Avenue, the old Bishop mansion, which became the first building in the world exclusively lit with electricity, causing frequent small fires. Edison himself moved to Gramercy Park not far from Cyrus Field. Upton remained behind at the light bulb factory, where the bulbs were being sealed with grain alcohol. So much of it was drunk by the dollar-a-day workmen that Edison tinted it green; any workman with green lips was then fired. Upton never worked at Edison's side again. He spent the rest of his life making light bulbs. The whole operation was slightly peculiar, at least in modern eyes, because Edison set up a separate company for each part of his business. But in 1880 the practice was not unknown. The bulb factory was personally owned by Edison and was licensed by, and sold its products to, the Edison Electric Light Company, which was owned by the Edison investors, including Morgan. The Edison Tube Works, manufacturing the underground conduits for the power system, was likewise owned by Edison and run by Kruesi. The Edison Machine Works on Goerck Street, another Edison entity, was run by Kruesi's old assistant, Charles Dean, and manufactured the dynamos. The Machine Works, but not the Tube Works, operated under license from the investors. The fourth Edison manufacturing entity, Messers Bergmann & Company, was a machine shop at 108–118 Wooster

Street, run by the German-born Johann Sigmund Bergmann, who had been Edison's off-premises machinist of choice during the Menlo Park years. Bergmann made Edison's consumer products — the chandeliers, the lamps, the sockets and plugs — and Edison had bought a third of the operation. The operation was complicated, it was run from under Edison's cap, and it was the sort of setup Morgan hated with every fiber of his being. But Morgan was a happy man right now. As promised, he owned the first electrical desk lamp in the world, and he even managed to hold his tongue when one of Edison's many short circuits burned his favorite desk to the ground. The damage was repaired by a young Canadian with the formidable name of Reginald Aubrey Fessenden, better known to the nicknaming Edison as Fezzy, who was trying to teach Edison something about insulation. Morgan had a steam engine in his backyard. The neighbors complained endlessly about the noise; Morgan ignored them. Sometimes the family neglected to tell the engine tender when they were entertaining, and the entire house went dark at eleven while they were still sitting over their port and walnuts. Morgan loved electricity, and for the moment he was quiet.

Morgan's installation was called an isolated plant, a self-contained electrical system in a private residence or a business, and to satisfy the huge demand for them, a fourth entity was established: the Edison Pioneers. The Pioneers built an isolated plant for Hinds, Ketcham & Company, a lithography concern, and for the first time the lithographers were able to do their work at night. They built an isolated plant in the basement of William Vanderbilt's new palace on Fifth Avenue, and it remained there until another short circuit set fire to Mrs. Vanderbilt's drapes; she discovered the steam engine in her basement and had hysterics. In the coming years, the Edison Pioneers were very busy indeed.

Batchelor was sent to Europe to arrange exhibitions in London and Paris. He never worked closely with Edison again. He died in 1910, toothless (probably as a result of the mercury poisoning he contracted as a result of his work on the Sprengel pump) and rich.

Grosvenor Lowrey was increasingly the lawyer representing the

Edison investors and less and less the lawyer of Edison himself. On March 23, 1880, he drew up a contract between them. The investors owned the digging license from the city, and the contract gave Edison permission to illuminate two portions of it. The first was "located in the lower part of the City of New York, bounded on the East by the East River, on the West by the middle line of Nassau Street, on the North by the middle line of Spruce Street, and on the South by the middle line of Wall Street." It was the very area Edison and his muckers had carefully surveyed during the previous year. But he seemed not to notice that he could no longer make a move without the permission of Morgan and the others, because although Edison owned pieces of all his new companies, the companies actually owned him, and the investors controlled the companies. The investors, to their horror, made the belated discovery that Edison was in the quack patent-medicine business. They ordered him to abandon it, and he did. To Edison, this exercise of raw muscle power suggested nothing.

The second, uptown location was to be designated later. The first was named the Pearl Street District. In the spring of 1881, Kruesi began to dig it with crews consisting mostly of Irish immigrants who knew nothing about electricity except that it emanated from a spirit, probably an evil one. Mistakes were made. And the makeshift insulation generated a gas. When exposed to a spark, it exploded. An entire portion of the largest city in the country was being turned into a bomb. Nothing could be done about it. New Yorkers, who had been so apprehensive about Morse's telegraph, had grown inured to electrical catastrophes, such as telegraph and telephone linemen cooking to death in plain sight. They might have been annoyed by Edison's explosive mains, but they weren't, taken as a whole, particularly frightened.

* * *

And there was one last breakthrough to add to the Mary Ann and the light bulb. It was Edison's soon-to-be-famous "feeder and main" system. The direct-current technology Edison used required him to taper the wires in his system, because a direct-current system was like a ditch: the more electricity was drawn off by the customers, the less electricity remained in the wire. At the dynamo, the diameter of the wires had to be large, because he was feeding a large amount of electricity into them. At the furthest reach of his system, wire diameter had to be small, because it was no longer carrying much electricity, and in between the two, the diameters gradually diminished. To meet the challenge, Edison decided to run large main electrical lines in all directions and feed them electricity at many points with smaller lines. This feeder-and-main system saved copper, because it resembled an interlocking system of larger and smaller canals rather than one big expensive canal that grew smaller as it went along. The solution, widely copied, was so simple and obvious that it was amazing that no one else had thought of it.

Edison's generators had a range of about half a mile. He always claimed that he bought the two adjacent buildings on Pearl Street because they were cheap. More likely, he bought them because they were in the geographic center of his franchise. The buildings were old, ugly, and badly in need of reinforcement if they were to support the heavy dynamos. The work was done. When the dynamos were installed and fired up, they began to hunt. As Edison shut them down, the rest of his men ran for their lives. Later, the armatures became so hot they had to be packed with ice.

Meanwhile, in London in 1881, other Edison men set up an exhibition in the Crystal Palace and installed a dynamo and a wiring system under the Holborn Viaduct near the City. London, too, had rules about digging up its streets, but using the viaduct meant they could avoid these laws. They lit 938 lamps, including 164 streetlamps at 32 candlepower. A portion of the General Post Office was lit, as were the Temple and various restaurants, hotels, and offices. Neither installation — the Crystal Palace or the Viaduct — was permanent, but Edison scared the living daylights out of the British government. The London *Daily News* wrote:

> There are two questions to solve besides the production of
> a lamp, viz: the proper distribution of electricity through a

town, and its economy relative to gas. Mr. Edison is far and away in advance of all rivals in the solution of these problems. His exhibition is the wonder of the show, and his representative is certainly the prince of all showmen. There is but one Edison and Johnson is his prophet. One feels after an hour with Mr. Johnson that there is nothing left to be done, that one's gas shares must be sold at once, that there is only one system, and that is Edison's and that every question had been solved.

Joseph Swan had finally laid hands on a Sprengel pump, achieved a vacuum, and made his bulb work, but Swan still had no distribution system; his lamps were still co-located — that is, they came equipped with their own batteries. The members of Parliament were faced with a stark choice: surrender to Edison, install his system and surrender the electrical market to an American inventor, or think of something else. They decided to think of something else, and they passed the Electric Lighting Act of 1882. It specified that any municipality wired for electricity had an absolute right to buy the system after twenty years. Remarkably few entrepreneurs were willing to take up a poisoned chalice of an offer like that. You would build the system, get twenty years to try to recover your costs and make a profit, and then the city could take it over and make the money that would rightfully be yours. This was not long-term thinking. For eighty years Britain had held a commanding lead in electrical theory and practice; as recently as the 1870s, the brilliant, doomed James Clerk Maxwell had written the equations that dominate the understanding of electricity to this day. Now, with the Electric Lighting Act, England became a vast electrical wasteland, its darkness pierced only by gas. The British never recovered their lead. But although the parliamentarians didn't know it, with the electricity law they had written Edison's death warrant in the electrical business.

On September 4, Edison put on a new Prince Albert coat, an unusual garment for a man whose clothes were usually stained with acid, and went to the Pearl Street station. The friend of the press was no longer talking to reporters. The system had been tested and retested. The problem on Nassau Street, where a portion of the pavement had been electrified, spooking every horse that stepped on it to the vast amusement of a growing crowd of spectators, had been repaired. But the Pearl Street grid was the largest and most complex thing Edison had ever done. Too much could go wrong, there were too many possible sleepers in the system, and if something failed, the negative publicity might be devastating. The Prince Albert coat did not remain immaculate for long; there was a dynamo problem at Pearl Street, and he fixed it while wearing his new garment. It hadn't occurred to him to take it off. At three o'clock, the appointed witching hour, he made his way to the Drexel Morgan offices at the corner of Wall and Broad, somewhat the worse for wear. In the Pearl Street station, his associates closed the switch. The lights in Morgan's office came on. The lights at the *New York Times* came on. Everywhere in the new power grid, there was light. There was no hitch. A remarkable event had occurred, and it seemed to happen almost effortlessly. And, like so many things with electricity, it happened silently, with no fanfare and no particular drama. The Industrial Revolution was incredibly noisy and incredibly disruptive. The Electric Revolution wasn't. The electric lights just came on. For the first time, people no longer had to burn things to illuminate their homes, and electrical devices no longer required batteries. The world would never be the same again.

And so Thomas Edison went forth to build out his system. Afterward, perhaps, he could rest on his laurels, maybe get back to a little inventing. But first, there was work to be done, a country, a world to light. He was, after all, the son of Canadian immigrants living up to his potential in a way that only America let some people do. But there was a certain irony here: Edison was about to be blindsided by another immigrant who kept trying to tell him he was using the wrong kind of electricity. And he was. He had, absolutely, built his system around the wrong kind of electricity.

CHAPTER EIGHT

THE MANUFACTURER
AND THE MAGUS

When an abnormal man can find such abnormal ways and means to make his name known all over the world with such rocket-like swiftness, and accumulate such wealth with such little real knowledge, a man that cannot solve a simple equation, I say, such a man is a genius — or let us use the more popular word — a wizard. So was Barnum! Edison is and always was a shrewd, witty business man without a soul, an electrical and mechanical jobber, who well understood how to "whoop things up," whose only ambition was to make money and to pose as a sort of fetish for great masses of people that possess only a popular notion of an art, and who are always ready to yap in astonishment at some fire-work display that is blown off for the benefit of mankind.

— *Francis Jehl*

The light bulb destroyed Menlo Park and it wrecked Edison as a major inventor. The years 1876–82 were his great years. Afterward, until the end of his long life in 1931 at the age of eighty-four, there was only one major invention, the alkaline storage battery, and it took him ten years to get it right. (There was the motion picture, but he used another's projector, so it was not a complete invention.) His wife, neglected and impractical, grown huge on a diet of chocolates and grown terrified with a pathological fear of rape, had gone insane and then died. Nikola Tesla, his former employee, and George Westinghouse, a Pittsburgh manufacturer, had defeated him in the electrical wars. Morgan took his company away. But uppermost in his mind was Tesla. Then Tesla was destroyed by Morgan too.

At a remove of a hundred and twenty years, it's a little hard to know what to make of Nikola Tesla. Nowadays, he would probably be medicated with psychotropic drugs into a permanent state of stun. We would never learn his name. But in the 1880s, with a single invention, he changed the entire world.

He was a Croatian Serb from the Military Frontier of the Austro-Hungarian Empire. Although he was born in modest circumstances to a father who was a Serbian Orthodox priest, he was a member in good standing of the Serbian professional class. One of his uncles was a professor of mathematics. Another was a field marshal. A young man with connections like that could go far, especially if he came from the Military Frontier. For the lads of the Military Frontier, there were generous scholarships. But Tesla was no ordinary lad. He was one of the strangest men who ever lived.

He was six and a half feet tall, weighed 140 pounds, and had long thumbs. This was thought significant because long thumbs supposedly indicated superior intelligence. But he has been poorly served by his biographers. One of them, Margaret Storm, writing in green ink in 1959, said that he was a Venusian who came to earth on the back of a gigantic dove. Another, Margaret Cheney, writing in 1981, believed in ghosts and Kirlian photography. He is variously credited with the invention of the laser, the ray gun, and the gigantic explosion that devastated Siberia in 1908. Others, mostly anonymous and dwelling on the Internet, claim that he lives on, deathlessly, with an equally undead Marconi as his roommate in an underground city in South America, where they occasionally nip over for tea with their neighbor, Adolf Hitler. Tesla, in life pretty strange himself, in death has become a magnet for every wingnut and squirrelcage in the known universe. Separating Tesla from the frequency noise that surrounds him is no easy task, and the picture that emerges is both weird and dim.

He was driven into a panic at the sight of a pearl earring. He required that his meals be served with seventeen napkins, so he could wipe the germs off his cutlery (he had looked through a microscope once and was horrified). Then he measured his food, dunked it in water to determine its volume, and divided it into three portions. Everything had to be done in threes. If he walked once around a block, he had to walk around it two more times. If he swam, he had to quit after twenty-seven laps. He believed he could communicate with pigeons. His favorite day of the week was Friday. His favorite number was thirteen.

Unlike other scientists, he rarely published in peer-reviewed journals. Instead, he preferred popular magazines. The *Century*, edited by one of his few close friends, was a favorite. In his writings he frequently sounded like a complete, and possibly dangerous, madman. He became the model for Dr. Zarkov, Flash Gordon's scientist sidekick. It was possible that he was also autistic.

He couldn't bear to be touched. He was intensely interested in himself, to the point where he told and retold childhood anecdotes of consuming interest only to himself. He believed that everybody, including himself, was an automaton and that thought was simply a reflection of external stimuli. These thoughts, he believed, could be both projected and recorded. Subject to periodic nervous breakdowns, he experienced periods in his life when the world was too much with him, and he was tormented by what he thought was the sound of a fly walking three rooms away. His eyesight, he claimed, was frequently obscured by brilliant rectangular shapes, and he was subject to visions. And he couldn't stop talking about all of the foregoing.

He claimed that his illiterate mother was a talented inventor in her own right, and he was convinced he had killed his older brother by causing a horse to rear when he shot it with a popgun. He was a brilliant student, and he was obsessed with alternating current, in part because it omitted the use of a commutator, a device he loathed. But not much of anybody else was interested in alternating current. Direct current, used by Edison, is the kind of electricity produced by a battery. It flows in only one direction, steadily. Alternating current, discovered by Henry and Faraday, is produced by a dynamo. It constantly reverses itself, and each pair of reversals constitutes a cycle. When Edison built his dynamos, he was obliged to equip them with the commutators Tesla hated, which abolished alternating current's cycle to achieve the direct current Edison preferred. The commutators sparked and wasted electricity. Direct current was weak. A direct-current central power station like Pearl Street could provide power to no more than a square mile, which meant that much of the countryside would never be electrified, and the copper wires grew ever larger the closer they came to the station. Given direct current's drawbacks, it is something of a mystery why Edison chose it, although he gave a number of explanations and told many lies. Perhaps he used it because it was the nearest thing at hand, it was well known, and alternating current wasn't. Whatever the reason, he was certainly determined to stick with it.

Tesla's alternating current, by comparison, was far more powerful. It could be transmitted over very long distances, although in the 1880s nobody knew how far. It could also be transmitted at extremely high voltages while using thin wires of a uniform diameter. Almost no one was interested in it except the Italians with their countless amps and volts of untapped hydropower in the Alps, a young Frenchman named Lucien Gaulard who was almost certainly insane, and the young Nikola Tesla. His professors told him that nothing would ever come of it. The world belonged to direct current. "Mr. Tesla may accomplish great things," one of his teachers told him, "but he will never do this."[1]

Reciting from *Faust* while walking though the City Park in Budapest after another of his breakdowns, he suddenly saw how he could make alternating current work. Tesla, like Edison, constantly adjusted the truth to fit his mood and audience, and although he came to America and became a citizen, he never understood Americans. Edison had him pegged there. But Edison was a very large part of the problem. Although Tesla almost certainly never had a sexual experience in his life (another thing he could not stand was human hair), all his life he was drawn, moth to the flame, to confident, assertive men. Edison was the first. When Edison told Tesla that he owed his genius to eating rarebit for breakfast, Tesla ate rarebit for breakfast for a month. And observing Edison through some dim romantic veil, he concluded that the scientist-inventor was a wizard, because he had invented so many things.

Charles Batchelor, losing his teeth and running Continental Edison out of Paris, almost certainly never wrote to Edison about Tesla, as some of Edison's biographers have him writing, "I have known two great men. You are one, and this young man is the other." Some months later, Batchelor would in fact write to Edison that the woods were full of Teslas. Nor was Elihu Thomson particularly impressed with the young Serb, but Thomson, who was almost certainly Tesla's inventive equal, was a conventional man who was doubtless put off by Tesla's mass of quirks and huge ego. Tesla could speak six languages and quote poetry by the yard. He was always immaculately dressed and worked in green suede hightops, cuffs and collars he discarded after a single wearing, frock coats, and white gloves. It was his ambition, he repeatedly said, to be the best-dressed man on Fifth Avenue. Although he was not a graduate of a European university, he was an accomplished billiards player. Although Edison's mansion contained a billiards room, Edison himself played precisely one game in his life, and he didn't finish it. Nikola Tesla was, in short, the sort of man Edison was inclined to loathe with every fiber of his being. Surprisingly, he didn't, not at first.

The path from the Dalmatian coast of Croatia to 65 Fifth Avenue was metaphorically a short one. In 1884, the year Tesla came to America, there were few people in the entire world who knew the faintest useful thing about electricity, and they tended to find each other. Through the good offices of his family, Tesla had gotten a job on the Budapest telephone exchange, only the fourth telephone exchange on the planet. But like all young electricians, Tesla wanted to work for Edison, and again through his well-connected family he was soon in Paris, where Batchelor took him on as a trouble-shooter. Anybody who knew something useful about electricity would do in a pinch, and pinch there was: the French loved Edison and the Germans were fascinated with him, and there was work enough for all. What happened next is more than a trifle murky, largely because Tesla is our only source, and Tesla quickly grasped the superficialities of our American myths, or at least the myths about inventors and their obligatory poverty-stricken origins. Be that as it may, he somehow found himself in New York with, he said, only four cents to his name. He made his way to 65 Fifth, and the rest — according to Tesla — is history. In any event, Edison took him on. Tesla seems to have spent his time working on arc lights, a business that didn't interest Edison very much, but one he halfheartedly entered because his investors saw the bonanza that was being reaped elsewhere by Davy's invention. Tesla's version of his work for Edison was rather different. In Tesla's version, he redesigned Edison's dynamos, and it was the cause of their falling-out. In Tesla's version, which he often retold and which appears in some of his biographies, he left Edison's employ because Edison had promised him $50,000 for making the dynamos run in synch and then reneged. In fact, Edison's dynamos were adjusted by hand, they never ran in synch, and only one out of the six of them was running at all. In Edison's version of the event, which he too retold, Tesla left because he asked for a seven-dollar raise and didn't get it.

The Tesla record is murky on this and many other points, not merely because Tesla was peculiar but because he had a remarkable visual memory. He claimed — with some justification, because unlike virtually all scientists he seldom took any notes — that he could see his inventions in his mind, run them, test them for balance, and watch them age, all in his head. From completed mental conception to finished product, he argued, nothing changed. But he also had a peculiarly Central European attitude toward the inventive process and the scientist's role in it; it all went back to Humphry Davy and the scientist as lone hero. Like Edison, Tesla had an exalted notion of the inventor's calling. He was secretive. His own lab assistants often had no idea what he was doing. He believed he was a member of the city's elite. He moved to the Waldorf as soon as he could afford it, stayed on long after he couldn't, and didn't pay his bills. He was served in the restaurant by the maitre d' himself, at a table reserved exclusively for his use, but he was not popular with the staff. In his lab, dressed in evening clothes, he entertained Mark Twain by passing hundreds of thousands of volts through his body, glowing and sparking in the dark, handling globules of pure electricity, and making mysterious tubes unconnected to a power source with wires light up. Once, he showed Twain a vibrating platform. Twain insisted on giving it a try. After a time, Tesla implored the great man to get off, but Twain, having a wonderful time, refused. Then Twain suddenly leapt off, demanded to know where the bathroom was, and bolted into it. As Tesla could have explained, the vibrating platform induced severe diarrhea. With his inventions, he expected to become fabulously wealthy, and he continued to expect it well into old age. He claimed that he could destroy the earth. He never became wealthy.

After leaving Edison, Tesla went into the arc light business in Rahway, New Jersey, backed by some of Edison's associates and a couple of local businessmen. Tesla, who even when young had an extraordinarily high opinion of himself, believed that his task was to build an improved arc lighting system, and he applied for his first patent. Then, when he'd returned the money due to his investors, he believed he could pursue his researches in alternating current. But Tesla's backers thought that his job after building the arc lighting system was to run it. Company with his backers was parted, Tesla was broke, and he claimed he kept body and soul together by working as a ditchdigger and planned to commit suicide on the midnight of his approaching thirtieth birthday. But given his phobias, even a short life of manual labor seems unlikely. It also seems too conveniently pat, especially in view of what he claimed happened next: that his foreman, hearing of his work in alternating current, put him in touch with a high-ranking officer in Western Union who had recently bested Jay Gould by building a competing telegraph system and forcing him to buy it. Possible, yes. But in the interest of verisimilitude — although nothing about Tesla in his strange life bears the stamp of verisimilitude as it is commonly understood — perhaps it's best to say that he obtained more backers, one of them a high-ranking Western Union official, and the new backers brought him to the attention of George Westinghouse.

* * *

George Westinghouse, a burly, athletic man with an enormous walrus moustache and a consuming horror of publicity, public speaking, and fame, was the next-to-last of the century's boy wonders. At the ripe age of twenty-two he had revolutionized the railroads with a remarkable new brake and a new signaling system that bore a close resemblance to a French optical telegraph. He also utterly defeated Thomas Alva Edison, not because he was a better inventor — he wasn't — but because he understood the distribution of natural gas through a man-made pipeline, another of his many ventures. Franklin had correctly surmised that electricity wasn't a fluid, but it behaved like one. So did natural gas. And an electrical system with a central power station resembled a railroad, with its main lines, branch lines, and switches. Understanding natural gas gave Westinghouse the greatest triumph of his life.

Alone among the electrical pioneers, Westinghouse grew up in a manufacturing environment. Morse's father had been a theologian and geographer, William Thomson's father had been a mathematician, and Edison's had been a jack-of-all-trades who turned his hand to everything from splitting shingles to running a grocery store. But Westinghouse's stern father built farming equipment in Schenectady, New York, where the lad had the run of the plant and had invented a pipe-cutting machine before the age of twenty.

In 1865, after spending the Civil War in both the army and the navy, he received the first of his 361 patents. It was for a rotary steam engine that, in practice, didn't work very well. He followed up with a device that returned derailed train cars to the tracks. And on a trip to Pittsburgh to buy steel for his new product, he discovered that a highballing railroad train was not an easy thing to stop. The only way to stop a train was to have a brakeman climb atop each car and turn a horizontal wheel that applied the shoes — a simple if time-consuming maneuver when the train was creeping into the station, but ineffectual if it was barreling along at full speed on the main line. The mortality rate among the brakemen was not encouraging. They fell between the cars. They collided with low bridges. In winter they slipped on the ice. Many were missing fingers. And, humanitarian considerations aside, the absurd railroad braking situation was costing everybody a great deal of money. The laws of physics came into play: although locomotives were growing ever more powerful, you could only have as long a train as the brakemen could stop. Fifty-car trains were possible but unheard of. Collisions were commonplace, even with telegraphic traffic management, which was essential. Accidents, sometimes horrible ones, were frequent, especially as the rolling stock improved and trains were now carrying quantities of hazardous materials, like oil from the Rockefeller fields in Pennsylvania. A major but unaddressed technological gap had appeared. But Westinghouse, a voracious reader like Edison, had encountered a magazine that described work on the Mont Cenis Tunnel in the Italian Alps.

Hand tools had advanced the work on the tunnel at a pace of eight inches a day. To finish the job in, as one observer remarked, "less than his lifetime," the chief engineer, Germain Sommeiller, built a reservoir high up on the mountainside that provided a head of water that compressed air. It worked like this: When water was poured into a pipe fitted with an airtight piston, it drove the piston down, and the piston compressed the air. The compressed air was then used to power mechanical drills, and the work moved forward twenty times faster. Remembering the article, Westinghouse devised a means to stop a railroad train quickly, with the same compressed air. Commodore Vanderbilt, the owner of the New York Central Railroad, was not amused.

"If I understand you, young man," the commodore intoned, "you propose to stop a railroad train with wind. I have no time to listen to such nonsense."

A persistent man, peddling his air brake at every opportunity, Westinghouse became known as Crazy George. Compressed air, like electricity, was not a concept that the average citizen easily grasped, and Westinghouse, in that fateful year of 1866, was only twenty years old. He was not merely crazy but a kid as well. The only person who took him seriously was a certain W. W. Card, the superintendent of the Steubenville division of the Panhandle Railroad. Card persuaded his superiors to give the Westinghouse air brake a test.

With the youthful inventor and railway officials embarked, the test train headed out of Pittsburgh toward Steubenville with a full head of steam. As it emerged from the Grant Tunnel just outside of town, a farm wagon was seen blocking the tracks two city blocks ahead. The farmer whipped up his team, the team reared, the farmer fell to the tracks, and the engineer applied his brakes. The train came to a halt when the farmer was still four feet from the cowcatcher. The moment was later immortalized in an historical painting, and Crazy George became rich. The nineteenth century was very fond of its miracles, and it was remarkably uncurious about its coincidences.

Because it was now possible to stop a train in something less than a mile, it was possible to reorganize the lines for vastly greater efficiency. Trains could become longer and longer as Westinghouse continued to improve his technology. More of them could be run down the same track and closer together. The telegraphers would still be needed to manage the traffic on the line, but they would not be desperately needed. New steam technologies could be rolled out faster, new locomotives and new parts for old ones, making everybody even more money. There were still workers called brakemen on the trains, but they no longer had to set the brakes. Thanks to Westinghouse, a railroad was no longer a road that ran on rails; it was an integrated, tautly managed transportation system.

Westinghouse perceived another niche for his technology. With more trains on the tracks, closer together, a new signaling system was needed. Westinghouse knew how to do it, using a mixture of compressed air and electricity to power the semaphores of the old optical telegraph systems. These semaphores would now become a means of communicating with the locomotive engineers. When the arm of the semaphore was up, the track was clear. When the arm was down, it was time to stop. When the arm was in the middle, the train was ordered to reduce speed and proceed with caution. As the signals further improved traffic management, more traffic would result, creating more customers for his brakes. George Westinghouse was set for life. The compressed-air part of the signals he could build himself, but to solve the electrical problems, he assembled a team. And, just as Tesla made his way to Edison in 1884 because the world of the electricians was still very small, Westinghouse's electricians had an excellent idea of what Edison was up to and how he was going about it. Some of Westinghouse's electrical people, like the formidably named Reginald Aubrey Fessenden, had recently been Edison men, and a few of them had some electrical ideas of their own that the increasingly rigid Old Man was becoming less and less willing to entertain. Westinghouse was not a rigid man; he listened to his people, and he treated them well. In an age of robber barons, he invented the half holiday on Saturday, and as his operations in Pittsburgh grew, the annual Christmas turkey became the first corporate pension fund. Westinghouse, like Edison, commanded a huge loyalty from his troops, and he was better than Edison at handling large numbers of them. When the Edison men in New York tried to unionize in 1886, Edison shut down his city shops and moved their operations to Schenectady. The Westinghouse shops, said no less a figure than the AF of L founder Samuel Gompers, could not be unionized, because nobody saw the point in joining. To his bankers' horror, George Westinghouse paid top dollar, built sanitary lunch-rooms and nurseries, hired college graduates for his labs, and supplied those labs with an unstinting hand. But although he was not a rigid man, he did have certain persistent quirks. He demanded speed; nothing moved fast enough for him, and to win a word of his praise his workers had been known to work on Sunday. He had a temper and he was high-strung. But nobody at Westinghouse knew the name of the operative who'd been fired for making a smart mistake. When hard times came in the nineties, a delegation from the shop floor came to Westinghouse's office and said the hands had voted. If it was necessary to keep the company afloat in the hard times, they would go on half wages. In his own and very different way, George Westinghouse was another of the century's remarkable men.

It was a funeral in Italy, a gas well, the British Electric Lighting Act of 1882, and the fact that Edison allegedly wouldn't give Tesla a seven-dollar raise that led Westinghouse to become the architect of Edison's undoing. If the Edison lab was organized like a machine shop, the Westinghouse lab resembled an art colony. There was credit grabbing and there was ego, and young William Stanley Jr. had most of the latter. A Yale dropout, he'd worked with Elihu Thomson on a silk-filament electric light; he'd helped Hiram Maxim, who went on to fame as the inventor of an improved machine gun, get his Edison-knockoff bulb up and running in a bank vault; he'd gone briefly into business for himself; and he became a Westinghouse hire in 1884 for the train-signal project. Westinghouse, always in search of the Next Big Thing, briefly considered founding an electric company of his own and putting Stanley's name on it, but he decided the time wasn't right. He had second thoughts after he gave his protégé Guido Pantaleoni permission to attend his father's funeral in Turin. Turin, as it happened, was holding one of the electrical exhibitions that were all the rage in Europe at the time. In the Alps, there were untold numbers of potential kilowatts of untapped hydroelectric power, and the untapped power had given the Italians a keen interest in alternating current. Alternating current, it was believed, could be conveyed over many miles, and the Alps were many miles away from the manufacturing centers and residents who would consume the electricity. Unfortunately, if the trick was to be done, the alternating current would have to be conveyed at very high voltages, and there was no way to step the current down. Induction coils, commonly called Ruhmkorff coils, like the one that had burned out the first Atlantic cable, stepped the current *up*, which was what was needed in a blazing arc light system. But no one had invented a transformer that would step the current *down*, which was wanted in an incandescent light system like Edison's. The Italians badly needed one, most European electrical scientists knew they did, and at the Turin electrical exposition, Pantaleoni found it on display. True, it wasn't a very good transformer and its inventor was not a likeable man, but there was nothing else like it. It owed its existence to the English Electric Lighting Act of 1882, the one that Edison had inspired, the one that retarded British electrical development for years.

That peculiar piece of legislation had an interesting feature. It was a common and much admired quality of British laws that they were often written vaguely, the better to be adjusted by the courts to the realities of the world. However, the Electric Lighting Act laid down a very specific requirement: that the owner of a power system could not dictate the wattage of the bulbs that were hooked into it. For British electrical equipment manufacturers, this created a nightmare. British bulbs came in wattages of anywhere from 46 to 91, creating a demand for electricity that varied with wild unpredictability. Without a predictable demand for electricity, nobody would build a grid, because the technology to vary the flow of electricity did not exist. It was here that the young Frenchman Lucien Gaulard saw his opportunity. With the aid of his British backer, John Gibbs, Gaulard built the world's first step-down transformer.

Because the transformer was designed for the strange lighting system mandated by the Electric Lighting Act, everybody else in the world regarded it as an extremely bad piece of work. Matters were not helped by the fact that Gaulard, who soon died in a madhouse, was almost as strange as the device he was designing. To everybody but the British, the transformer was needlessly complicated, with an armature that could be cranked in and out as the station operator tried to adjust the current to the line, an inexact procedure that gave the transformer too many moving parts that could break. It was a transformer that no one but a Briton could possibly love, and one did. He was Sir Coutts Lindsay, proprietor of the Grosvenor Gallery, the gallery of the "greenery-yallery" aesthetes satirized by William Gilbert in *Patience*, the gallery where the American impressionist James McNeill Whistler hung his controversial paintings. To Sir Coutts, the arrival of Edison's electric light was a godsend, because gaslight smoked and smudged paintings, and he immediately installed an isolated plant to light his gallery. Then a bug seemed to bite him; he became interested in electricity, and he began to attend electrical exhibitions, where electricians brought their wares to the attention of the public. At one such exhibition he saw the Gaulard and Gibbs transformer, and he was smitten. It was not a perfect answer to the power management problem caused by Britain's variable wattages, but it was a usable stopgap and it showed the way to an even better solution. Sir Coutts bought it, and he brought Gaulard and Gibbs into his electrical operation. Able to manage his power, although not well, Sir Coutts Lindsay became Britain's only electrical tycoon, and the Grosvenor Gallery became the only central power station in the kingdom.

Now Gaulard and Gibbs were in Turin, because they were men with a desperate need and the Italians were an obvious market for their device. Sir Coutts had hired a young electrical genius named Sebastian Ziani de Ferranti, who didn't like the Gaulard and Gibbs transformer and was kicking them out of the gallery, and elsewhere their patent was being challenged. Gibbs, losing his fortune by defending it, was in the process of appealing adverse court decisions all the way to the House of Lords. In short, Gaulard and Gibbs were running out of money, and the Turin exhibition not only offered them exposure in an obvious marketplace at a critical time, but the exhibition would award a grand prize of 10,000 badly needed francs. Pantaleoni, taking time out from the events surrounding his father's funeral, dropped by the exhibit hall, saw the transformer, and dropped Westinghouse a line.

Westinghouse, a man of boundless energy who frequently interrupted his meal to borrow a pencil and scribble down an idea, was off on a new tear. He'd drilled an amazingly productive gas well in his wife's flowerbed at their new suburban estate, Solitude. Now he was proceeding to supply the citizens of Pittsburgh with gas, but, as Westinghouse knew, natural gas differed from gas manufactured from coal. You could control the pressure of manufactured gas at its point of origin, but you could not control the origin pressure of natural gas, and the pressure at Westinghouse's wellhead was very high. When Pantaleoni's message reached him, he was developing a device to step down the pressure before gas was delivered to the customers, lest the houses of Pittsburgh be blown away. Now Pantaleoni was telling him that a device existed that could step down alternating current, much as his new invention stepped down the pressure of the gas from his well. Perhaps there was something in this, after all. He instructed Pantaleoni to buy the American rights to the Gaulard and Gibbs apparatus. Later, in 1886, he bought a couple of their transformers and gave them to Stanley to break down and analyze.

Pantaleoni wasn't the only person to examine the Gaulard and Gibbs machine at the Turin exposition. A Hungarian engineer named Otto Titus Blathy, who worked for Ganz & Company in Budapest, also took an interest — because, he later said, the Gaulard and Gibbs showed him how *not* to design a transformer. Apparently he learned a great deal. Soon he and two colleagues had designed their own transformer, the ZBD, with two results. First, Ganz became one of the principal European manufacturers of alternating-current power systems. Second, Edison bought the American rights to the ZBD. As he had repeatedly told Tesla, he had no interest whatsoever in alternating current. Indeed, he would soon come to hate alternating current. But the ZBD transformer was the best one on the market, and buying the American rights meant that it would not fall into the wrong hands. By owning the American rights, Edison could prevent any American from obtaining a ZBD transformer and setting up an alternating-current power system, by simply refusing to sell it. At that point, he had no idea that Westinghouse had William Stanley reverse-engineering the Gaulard and Gibbs transformer.

From this distance in time, only two things are clear about what happened next in Westinghouse's laboratory of prima donnas: Stanley, who had worked with Elihu Thomson, an early advocate of alternating current, didn't think much of the Gaulard and Gibbs transformer, and he began to have trouble with his colleagues.

Westinghouse's partisans claim that it was only through Westinghouse's efforts that the transformer project was pushed to a successful conclusion. It's true that Westinghouse had ordered Pantaleoni to secure the rights, had ordered the Gaulard and Gibbs machines and given them to Stanley, and had sent Pantaleoni to discuss the project with Werner von Siemens, the great German electrician. But Westinghouse seldom wrote letters, kept no journal, rarely gave interviews, and never told his side of the story. Stanley did. According to Stanley, it was only through the heroic efforts of Stanley that George Westinghouse was dragged, kicking and screaming, into the electrical business. In any event, it was deemed prudent and in the best interests of all concerned for Stanley to go and build Westinghouse's transformer somewhere else. Some say he had worked himself to the point of a nervous collapse. Others say it was for his own good. Perhaps both versions are correct.

Stanley moved his operations to Great Barrington, Massachusetts, where he had summered as a child. It is probably significant that he had with him a quantity of Westinghouse's money. And, in short order, Stanley not only had developed an improved transformer but had wired Great Barrington for alternating current. The line, stapled to trees, was four thousand feet long, nearly twice the reach of a direct-current central station. The current flowed at 3,000 volts. The transformer stepped it down to 500.

The Gaulard and Gibbs transformer had fed a power system that was wired in series. All high-voltage power lines were wired thus, and virtually all of them fed arc lights, which needed high voltages to produce their brilliant illumination. A system wired in series is like a string of Christmas tree lights: when one component of the system burns out or is shut down, the entire system shuts down. Charles Brush got around the series problem by equipping his system with a number of induction coils, which isolated the offending light without changing the system configurations; Elihu Thomson, now operating out of Lynn, Massachusetts, benefiting from the deep pockets of the local shoe manufacturers, and flying the flag of Thomson-Houston, did the same. Thomson-Houston soon bought out Brush, and Thomson was also moving in the direction of alternating current. He was delaying only until he had invented two safety devices: the lightning arrestor, a sort of lightning rod for high-tension power lines, and the magnetic blowout switch, which shut the system down if a power surge occurred. Westinghouse as yet had no safety devices, but he pressed forward nonetheless. Parallel wiring, isolating the lines running into the customers' homes to make each individual circuit behave as though it were directly connected to the power station — which Edison also used as a way of getting around the series problem, running individual lines from his mains to the homes and businesses of his customers, so that the wiring diagram of his system resembled a ladder — insured that the whole system remained up and running no matter how many of its components were down. This was an essential feature of an incandescent lighting grid where the customers click their lights on and off according to no easily predictable pattern. In a parallel system, a customer could flip a switch without turning the whole city off. It was a solution that used a lot of wire, it became the preferred way of wiring a power system, and it didn't need a transformer. Wedded to direct current, Edison didn't want people to start thinking about transformers. Thinking about transformers could lead to thoughts of alternating current. And now the worst had happened. In Lynn, Massachusetts, not far from Edison's flawed installation in Great Barrington, Stanley had created his own parallel wiring, and he was running Westinghouse's alternating current through it at 133 cycles per second, a speed that kept the lights of the customers from flickering.

For Westinghouse, two tasks now remained. First, it was essential to prove that alternating current could indeed be transported over long distances through economically thin wires. Second, someone had to invent an electric motor that ran on alternating current. Designing a motor for direct current was easy: in all essential respects, a direct-current motor was simply a dynamo running backwards. But many strange things happened in an alternating-current system. For one thing, there was lag: an electrical effect was observed shortly *after* a switch was thrown, not instantaneously. But alternating-current motors running on high voltages would, in theory, be immensely powerful, whereas direct-current motors were weak. A simple law of physics was involved: a motor running on a higher voltage did more work. An alternating-current electric motor could usher in an era of industrial productivity the likes of which the world had never seen. But this kind of motor was possible only in theory. No one could figure out how to invent one. Alternating current, said Edison in almost the exact words the British parliamentary committee had directed at his electric light, "is not worth the attention of practical men." Then Nikola Tesla gave a speech in New York.

* * *

For once in his life, it is possible to describe Tesla's activities in reliable detail, because he had backers who left records, and his backers had no reason to embellish the truth. They were Charles Peck, a lawyer who was a family friend of William Stanley, and Alfred Brown, a Western Union superintendent, both of whom had recently separated Jay Gould from a pleasing sum of money and were eager to put it to work in a new play. While he was supposedly working as a ditchdigger, Tesla had found the time and money to patent a thermomagnetic motor that ran on magnetism created by heat (because heat improves a magnetic field), and it was a good thing he had. The thermomagnetic motor didn't work very well, but if Tesla had brought Peck and Brown an alternating-current motor in the year 1885, it's likely they wouldn't have given him the time of day. Westinghouse and Elihu Thomson hadn't launched their power grids yet, and the future of the world belonged to direct current. Thomas Edison said so, and Edison had the ears of his countrymen. Peck and Brown rented Tesla a lab in lower Manhattan, they imported an old Tesla sidekick, Antony Szigety, from Europe and made him Tesla's assistant — he always worked best when he had someone to admire him — and they also retained the services of Parker Page, a patent lawyer and the son of Charles Grafton Page, inventor of the Page relay and the battery-driven train. Tesla himself was paid $250 a month. Then Tesla told them what he really wanted to do — build a motor that ran on alternating current — and the deal, which the backers thought centered on a thermomagnetic motor, almost fell apart.

In Budapest's City Park, at least as Tesla told the story, the vision of a polyphase — Tesla's word for multiple magnetic fields — alternating motor had come to him whole and entire, and he hadn't changed it since by a jot or a tittle. Polyphase, as Tesla conceived it, created rotation in the motor with two or more rotating magnetic fields that were out of phase with each other; when one magnetic field stopped, the next one took over. The result was a fast and clean little device with no sparking and hard-to-adjust commutator, and it was instantly reversible. Or so it would be when he built it. Tesla, who had the instincts of a master showman and later became one, proposed a bargain to his extremely reluctant patrons. If he could duplicate the egg of Columbus — the egg Columbus cracked and stood on its end, persuading Queen Isabella to pawn her jewels and bankroll his voyage — would they give him a budget to develop his device? Tesla was usually good at getting money out of potential backers the first time he tried, until they found out what he actually did with it, which was usually something different from the plan his investors believed they were backing. But here there was no question of what he was going to do. He had been trying to build his motor for years, ever since the vision came to him in Budapest. Peck and Brown agreed to the egg experiment.

Tesla obtained an egg and had it copper-plated. Then he placed a concealed electromagnet under his tabletop. When his patrons next visited the lab, he turned on the magnet. The egg stood up. Then, demonstrating the principles of his motor, it began to spin. But the few Tesla historians who have managed to winkle this anecdote out of the sparse Tesla archive may be making too much of it. Peck and Brown were hard-minded men who had defeated Jay Gould, and not many people had ever done that. It is unlikely that they would be won over by a party trick. It is far more likely that they had noticed something: George Westinghouse had gone into the electricity business, and he didn't have a motor. If Tesla succeeded, they would. A market would exist, and they would own it.

As Tesla proceeded, Brown noticed a problem. Tesla was in love with the two-phase motor he built, but it needed four wires. The systems being rolled out by Westinghouse and Elihu Thomson had only two wires. Later in life, Tesla was not good at taking orders even from the bold and forceful men he so admired, but the Tesla of 1888 was in no position to dicker because he was in no position to write another meal ticket. In 1888, it was Brown's way or the highway. In a few weeks, Tesla had produced a split-phase — the next century's word for Tesla's preferred "polyphase" — motor with two wires; he had introduced an induction coil and a resistor into his circuit, which created two out-of-phase currents that produced rotation. It wasn't elegant, for in later years he almost never mentioned it, but it worked. For the backers, it was show time.

Peck and Brown intended to build Tesla slowly. First, they had to establish his credentials. Nobody except a handful of people around Edison knew who Tesla was, and some of those probably didn't even know his real name, because Edison, with his compulsion for nicknames, had called him "our Parisian." Tesla had joined no professional societies. He had no social life — he was the one person Edison ever met who could equal or exceed his work schedule — and like Edison he preferred to work at night and alone.

In building Tesla, Peck and Brown succeeded beyond their wildest dreams. All they knew for certain was that they had a motor that worked on alternating current. Any scientist with vision would realize this motor was one of the great inventions of history, and Peck and Brown found just such a scientist. He was William Anthony, a Yale man who had become professor of physics at Ezra Cornell's university and was now chief engineer of the Mather Electric Company in Manchester, Connecticut. He was ideal. He had impeccable academic credentials, and he had functioned as a practical man of affairs in the real world. Tesla's motors deeply impressed him.

"I [have] seen a system of alternating current motors in N.Y.," he wrote to a colleague at the University of Wisconsin, "that promised great things." He also spoke warmly of it at MIT. Overnight, Tesla had achieved academic respectability. Next, Peck and Brown approached the trade press. Just before the patents were issued, they called in Charles Price of the *Electrical Review* and the shaven-headed, heroically moustached Thomas Commerford Martin of *Electrical World*. Both were favorably impressed. Martin and Anthony were even more impressed than they let on, and they simultaneously happened to be serving as vice presidents of the American Institute of Electrical Engineers, the most august professional body in the land. They persuaded Tesla, after some initial reluctance, to speak to the membership. Martin, who usually chaired the meetings, relinquished his position for the night so he could ask questions from the floor — a sure sign to the other members that something was up. Although Tesla had the flu and waited until the night before the meeting to write his speech, he proved to be an accomplished public speaker, with a fine sense for the natural balance of an English sentence and a scientific clarity that was a model of its kind. He described his motor. His audience was stunned.

Perhaps not even Tesla knew what he had done; he knew next to nothing about economics, as the rest of his life abundantly demonstrated. But his audience knew that Nikola Tesla, a man of whom none of them had ever heard, had invented tomorrow. Tesla's motor changed everything. Its underlying principles showed how to build a radical new power system. "In the steam age," wrote Matthew Josephson, one of Edison's more perceptive biographers,

> the machine was a superbeast of burden requiring the physical degradation of men for its operation; the tasks accomplished were primarily those that could have been achieved by a sufficient quantity of manpower; most machines were but an extension of man's larger muscles. Early in the electrical age, on the other hand . . . the view of the machine as a benevolent superhuman which increased leisure and physical comfort became more

compelling and more explicit. Machines began to do things that no quantity of men could do.[2]

Tesla had always had a good opinion of himself. Now he discovered that he was a great man. In London he addressed the Royal Society. There, he sat in the great Faraday's chair and was given the great Faraday's whiskey to drink, a very great privilege.

But the future wasn't going to arrive anytime soon. First, Westinghouse had to be persuaded to buy the motor. Then he had to figure out what to do with it. He wasn't out to change the world. He was out to make a buck. As were Tesla's patrons. The idea was to start a bidding war and jack up the price. Charles Coffin, the ex–shoe manufacturer who ran the business side of Thomson-Houston, was approached. He flipped the ball into Elihu Thomson's court. Thomson was working on his own alternating-current motor, which, he was certain, would be superior in every respect, and he was opposed to buying outside patents. Thomson-Houston passed. Westinghouse remained intrigued, especially if (as he may have surmised) Thomson was up to something on his own. Oliver Shallenberger, one of his engineers, had recently dropped a small spring on an alternating-current arc light. The spring began to rotate. It was an old electromagnetic effect, one observed by both Faraday and Henry, but Shallenberger had never seen it before. He told Westinghouse that, using this effect, he could design either a meter or a motor. Westinghouse badly needed a meter to measure the electrical consumption of his future customers, and he told Shallenberger to go ahead with one. Now Tesla's patrons were proposing to save him time by offering a motor he wouldn't have to build from scratch.

Westinghouse had another thing on his mind. Studying Tesla's patents, he began to suspect that they covered *every* rotating-field induction motor. If that was the case, he was confronted with only two choices. He could let Shallenberger build his motor and pay Tesla a royalty, or he could buy Tesla's patents. Before he made his decision, he had to learn how good the Tesla motor was, and he sent two of his executives to New York. They recommended that he buy the motor, but the people around Tesla were demanding an outrageous price: more than $200,000 and $2.50 for every horsepower produced. They claimed they already had an offer at that level. Anybody who bought the Tesla motor would end up paying Tesla and his people billions. Westinghouse then dispatched his star inventors, Shallenberger and Stanley, to look at Tesla's motor. Shallenberger grimly told Westinghouse to buy it no matter what the price. Westinghouse met the Tesla terms. Everybody was going to get very, very rich.

But not anytime soon. As mentioned, Westinghouse's power plants ran at 133 cycles. Tesla's motor ran at 60 cycles. The two were completely incompatible. Tesla agreed to come to Pittsburgh and work on the problem. Peck and Brown somehow dropped out of the picture and were never seen again.

In Pittsburgh, Tesla began the life of luxury-hotel living that he never afterward abandoned. By nature a solitary man, he did not work and play well with others, especially now that he was a certified genius. He may, however, have worked for free. As a genius, his services were something to be bestowed, not bought. Tesla became deeply attached, emotionally, to George Westinghouse — another of his patterns. He was never able to resolve the problem with his motor, the first of many noncompletions that would characterize the rest of his long life. Tesla, despite his original insistence that a few trifling modifications would bring success, could not make his motor run on Westinghouse's 133 cycles, and Westinghouse could not be expected to retrofit his 120 existing electrical plants. Perhaps to preserve everybody's peace of mind, Westinghouse soon sent Tesla home. He had another problem. Edison was attacking.

* * *

Edison had become an old man while young, and he was never a young man again. In 1886 he tried to return to inventing and built a splendid new lab in West Orange, New Jersey, where he set out to perfect the phonograph he had abandoned a decade before in his quest of light bulbs and power systems. That part of his life, he thought, could run itself now. Still a master of publicity, he announced a phonograph vigil where he and his associates would labor sleeplessly for days until the task was done. He capped the experience by posing for one of his most famous photographs: Edison slumps exhausted, rumpled, and resembling, as Sarah Bernhardt had once remarked, the first Napoleon in one of his brooding moods as he displays the perfected talking machine. Actually, the work had gone easily, if not quietly. The new lab was almost a parody of the Edison method at its haphazard worst. He tried to gather a sample of every substance in the world — "from a walrus hide to the eyeballs of a U.S. senator," he boasted. There was a sample of every screw in the world, and of every chemical. If Edison, in a fury of invention, wanted to lay hands on something, he had only to go downstairs. Like Menlo Park, the lab's activities centered on a single man, Edison himself. If the industrial research lab was being born anywhere, it was being born in the Westinghouse shops in Pittsburgh, where teamwork and scientific discipline were the order of the day, and in the Edison shops in Schenectady, where he had moved them from Manhattan to escape the unions and had placed the orderly, methodical Kruesi in charge while keeping his own hands off. But in 1888, when it was completed, Edison's fully equipped lab was very new. Upton and Batchelor were still around, Batchelor in a much reduced role and Upton running the light bulb factory in New Jersey, but they gradually drifted out of his life. Edison's new team had not assembled, and he threw the lab open to any fellow inventor who wanted to try out the facilities, which were the most advanced in the world. It was a mistake. The West Orange lab was immediately deluged with every crank and teenage male in the Northeast. Explosions occurred. Perpetual motion machines were assembled. One inventive group was toying dangerously with chlorine gas for some reason or another, and they kept inhaling it; ether was a partial antidote, and the halls were littered with somnolent experimenters. Meanwhile, in another part of the lab, Edison was electrocuting cats and dogs.

Some of his associates begged him to throw in the towel and get into alternating current. Its virtues were obvious and the drawbacks of direct current — the weak electricity, and the fact that you had to build another power plant every half mile — seemed insurmountable. But a hitherto unsuspected dark side of Edison's personality had surfaced. He had become obsessed with defeating Westinghouse and alternating current, largely, it seemed, because it hadn't occurred to him to use it. He was convinced of his genius, he used direct current, and it therefore followed that a genius such as himself used nothing but direct current. It wasn't very clear thinking, but there was no talking him out of it. His inventive objective, he once told a reporter, was not so much to make money as it was to beat up on the other guy. In electricity, the other guy bore the hydra heads of Tesla, Westinghouse, and Elihu Thomson. Edison went straight for the jugular.

When attacking the gas companies in the interest of stealing their customers for his electric light, he placed great emphasis on the unhealthy side effects of gas lighting. A number of these effects did not exist. Gaslight did not, for example, cause shortsightedness, as Edison claimed. Now he tried the same tactic on the alternating-current crowd: he badmouthed their electricity, making claims that sometimes did not bear close scrutiny. The Edison company got up a handsome gold-bound pamphlet entitled "A Warning," in which Westinghouse and Thomson-Houston were called "patent pirates." Simultaneously, an electrical engineer named Harold Brown took up the cudgels on Edison's behalf.

So thoroughly has Harold Brown dropped out of the pages of history that it's no longer clear just who he was. At some point, he seems to have taken lab space in West Orange. He may have handled certain patents for Edison in Chicago. He may have sold Edison's electric pens. He does not seem to have been an employee of Edison's, but what he did for a living is uncertain. Clear beyond a shadow of a doubt, however, is the fact that Harold Brown hated alternating current with every fiber of his being, and Edison was more than willing to have him hang around. Brown worked up a list that claimed to name thirty people who had been electrocuted by the dangerously high voltage of alternating current. Westinghouse's investigators found that in twelve of the relevant cities there were no Westinghouse plants at the time of the deaths. Of his list of thirty, one death was caused by drowning, sixteen deaths were caused by direct current, one was caused by alternating current, and in the rest of the cases no death had occurred. Harold Brown, effectively rebutted, decided to electrocute some animals. His increasingly desperate counterattack had three interesting results: the electric chair, the end of the country's seemingly irresistible anti–capital punishment movement, and Herman Melville's great novel *Billy Budd*. "Just as certain as death," Edison intoned,

> Westinghouse will kill a customer within six months after he puts in a system of any size. . . . There is no plea which will justify the use of high-tension and alternating currents, either in a scientific or a commercial sense. They are employed solely to reduce investment in copper wire and real estate.[3]

In this remark, although there was some truth to it, there was the germ of a very bad idea. At the very moment Edison uttered it, New York State was locked in an intense debate on capital punishment — whether to do it at all, how to do it if it was to be done, and how not to do it. Concerning the last point, there was general agreement: hanging was to end in New York. Sometimes the victims slowly strangled to death, sometimes they were decapitated, and particular attention was paid to the fact that the victims were often drunk. "In the moral aspect," remarked one observer, "the gross impropriety of sending a man to his Maker intoxicated is too obvious to require comment." The public also objected to hanging women, although the portion of the public that favored capital punishment had no objection to killing them in some other way. The public, however, was far from unanimous on the subject. Indeed, a law abolishing all forms of capital punishment in New York had failed in the legislature by a single vote, and it seemed only a matter of time before the abolitionists carried the day. Searching for a more humane method of slaughtering criminals, the governor convened a "death commission" to study the matter and make recommendations. In no time at all, the commission rejected the idea of burning criminals at the stake.

The commissioners were three. Elbridge Gerry, the chairman, was the grandson of the vice president of the United States who signed the Declaration of Independence and invented the gerrymander. A wealthy man, he was commodore of the New York Yacht Club, the first legal counsel to the American Society for the Prevention of Cruelty to Animals, and a founder of the Society for the Prevention of Cruelty to Children, which at one time was called the Gerry Society. Like Morgan, he loved electrical devices and named his yacht *Electra*. The second member, Dr. Edward Southwick of Buffalo, the commission's most outspoken member, was one of the most skilled oral surgeons of his time. Known as Old Electricity, he experimented on animals to discover how much electricity it took to kill them, and he had once seen a drunk killed by a live wire. The electricity in the wire was direct current, and the drunk seemed to die painlessly. Franklin, for one, could have told him otherwise. Southwick later became known as the Father of Electrocution. The third member, Matthew Hale, was an Albany jurist.

Although the cards were clearly stacked, the commissioners eventually put forward four conclusions. The guillotine, they concluded, was "the most merciful" method of execution, but it was also "the most terrible to witness." There was a danger that it would inspire bloodlust in the spectators, and it was French. The garrote was Spanish, un-American, and medieval. Of prussic acid they had little to say. Electricity, on the other hand, was humane and painless. Needless to say, electricity carried the day in the state legislature. The governor signed the bill mandating the electric chair on June 4, 1888. On June 5, Brown's first letter denouncing alternating current appeared in the *New York Post*. It was all a big coincidence. Another big coincidence was the fact that the French had recently blown the price of copper into the stratosphere, placing Edison's fat wires at a distinct competitive disadvantage to Westinghouse's thin ones. If Westinghouse got a commercial alternating-current system up and running, Edison's home court advantage was lost. In addition to his pride, his pocketbook was threatened. For sound business reasons, he had to head off alternating current.

There was a brief statewide debate about just what the new humane method of execution would be called. Chairman Gerry favored *electrolethe*. The *Scientific American* came forward with *thanelectrize*, *fulmenvoltacuss*, and *electropoenize*. Edison, always quick to coin a phrase, suggested that murderers be "westinghoused" in the "westinghouse." Southwick thought "electrocution" would do, and electrocution it was, although as Brown's campaign gathered steam in the coming months, somebody suggested "browned."

And Brown's campaign did gather steam. Edison tried to obtain stray animals from the Humane Society, and when it refused, he offered twenty-five cents to the street urchins of West Orange for any stray animals they might find. Soon the dogs and cats of West Orange began to disappear, but Edison was quickly put out of the cat business, because they were too hard to electrocute. Batchelor gained firsthand experience with the painless new method of execution when, trying to strap a lively mutt into a makeshift electric chair, he gave himself a good jolt. "I had the awful memory of body and soul being wrenched asunder," he wrote, "the sensation of an immense file being thrust through the quivering fibers of my body." But Edison's and Batchelor's experiments were conducted in privacy for the benefit of Edison's friends, the reporters. Meanwhile, secretly and through his agents, Edison began to lobby for a New York law that would have limited alternating transmission to 300 volts, destroying Westinghouse's huge economic advantage, because 300 volts wasn't enough electricity to make any serious money from; it wouldn't light up many bulbs, and nobody could run a factory with it. Brown was Edison's public face.

Before a live and horrified audience at the Columbia School of Mines, Brown first tormented a dog with direct current, showing that DC was uncomfortable but harmless, and then killed it with AC. He wanted to electrocute a second dog, but the audience stopped him. Sportingly, he challenged Westinghouse to a duel. Brown would take a healthy dose of direct current into his body, confident of the result. Westinghouse, so fond of lethal voltage, was invited to do likewise with his current of choice. Westinghouse declined. Brown crowed. When it was pointed out that a human was much larger than a dog, he electrocuted a calf. Then, for good measure, he electrocuted a horse. Years later, when the dust of battle had settled but not Edison's bitterness over the outcome, Edison electrocuted an elephant.

An appalled Westinghouse refused to sell an alternating-current dynamo to New York and, it was widely believed, gave funds for the defense of William Kemmler, the ax murderer selected for the experiment. Edison and New York State employed front men to buy an AC dynamo anyway. On August 6, 1890, Kemmler was strapped into the chair at Auburn Prison. The first jolt lasted seventeen seconds and failed to kill him. The second application of current lasted much longer. Kemmler began to cook. A reporter fainted.

"Strong Men Fell Like Logs!" the *Tribune* headlined the next day. The formerly pro-execution *Times* was only slightly more judicious. Kemmler's execution, the paper editorialized, was "a disgrace to civilization."

"They could have done better with an ax," said Westinghouse. But New York State was delighted and so was Edison. Kemmler, Edison proclaimed triumphantly, had been "westinghoused" in the "westinghouse," and the proclaimed dangers of alternating current were clear for all to see. But if the advantage to Edison's anti-Westinghouse campaign was clear, the advantage to the nation was not. Kemmler had died badly and probably in immense pain, and the literature describing the unpleasant sensations attending an electrical shock was as old as Franklin. Electrocution was not a merciful way of killing someone. Still, except in those states that clung to hanging or changed to the equally unmerciful gas chamber, America took the electrocution of criminals to its heart. Nationwide, the anti–capital punishment movement lost its bearings for the next eighty years. In New York City, an interested observer named Herman Melville wrote his last novel. *Billy Budd, Foretop-man* is many things, but it is primarily a tract on capital punishment.

In the end, it was all for nothing. Westinghouse's alternating current carried the day because it was the best way to carry electricity to the largest number of people. He proved it in Chicago, and then he proved it again at Niagara Falls. In the immediate short run, Edison continued to improve and expand his direct-current systems, as did other direct-current companies such as Fort Wayne Electric and anything with Edison's name in it, such as Commonwealth Edison. The latter were in the same bind that Westinghouse faced when he attempted to introduce Tesla's motor and adapt his dynamos to polyphase operation. The installed base of DC stations and power lines was simply too new and too expensive to convert to superior AC electricity.

Still, science marched on — everywhere, that is, except in Britain. For five years the Grosvenor Gallery remained the only power station in England. The British did no work in polyphase motors or dynamos, and the European baton of electrical progress passed to the Germans and, to a lesser extent, the Hungarians. There were many complaints that Americans had seized the lead, that the Americans had taken a European technology — the alternating current, the transformer — and run with it while the Europeans had ignored their own breakthrough achievements.

And the Americans were certainly running. Elihu Thomson, having satisfied himself on the safety issues, was moving Thomson-Houston into AC. A researcher named Charles S. Brady invented the rotary converter, an electromechanical device that was simultaneously a generator and a motor that allowed a DC system to be fed by AC power lines. It did more: it also allowed a polyphase dynamo working at 60 cycles to be connected with the old single-phase 133-cycle AC systems. Universal power — power that ran both motors and light, power that ran on AC or DC systems — was at last possible. At Telluride, Colorado, George Westinghouse strung a two-mile AC line. In Oregon he strung a four-mile line. The theory seemed to be true: AC could project electricity over lines for some distance, although four miles still wasn't much of a distance.

Then the Europeans stole a march. In 1890, to settle whether the city of Frankfurt would build an AC or a DC system, an electrical exhibition was held. The Germans, using polyphase equipment and the surplus electrical power of a cement factory, brought power to the festivities from 108 miles away. It was, by an order of magnitude, the longest transmission of electrical power in history, a stunning achievement that DC electricity could never match.

* * *

By the 1890s, Chicago was known as the Black City, for the filth of its industries, and the Red City, for its history of violent labor strife. And to an America still struggling to adjust to a huge wave of non-British immigration, Chicago was extremely uncomfortable for another reason: 78 percent of its population was foreign-born. "Having seen [Chicago], I urgently desire never to see it again," said Rudyard Kipling in 1889. "It is inhabited by savages." To clean up its act and to revise its image, Chicago set forth to hold the World's Columbian Exposition. It was a world's fair like no other, commemorating (in 1893, a year late) Columbus's discovery of America. During the exposition, Chicago itself was renamed the Windy City, a distinct improvement.

Although the architects Louis Sullivan and Frank Lloyd Wright regarded the design of the exposition's centerpiece, the White City, as an unforgivable betrayal of the new architecture being developed in Chicago, the rival architect Charles Burnham's guidelines laid the ground for the City Beautiful movement that would dominate American urban thinking for the next three-quarters of a century. With its uniform sixty-foot cornice line, its buildings in the French Beaux-Arts fashion, broad avenues, and reflecting pools, it cut the template for the Mall in Washington, countless empty urban plazas, and every government building erected into the 1970s.

The exposition's Midway Plaisance, a distinctly guilty pleasure featuring Little Egypt and the world's first and largest Ferris wheel, was the model for every carnival thereafter. The Rumford Kitchen, staffed by local socialites, tried to bring the miracle of modern cookery to the American housewife. Cracker Jack, Aunt Jemima syrup, Cream of Wheat, shredded wheat, Pabst beer, and Juicy Fruit gum were introduced to an eager public. So was grapefruit, an exotic new breakfast food. Edison had promised an exhibition of his new kinetograph, the motion picture machine that was his last world-changing invention. Typically, most of the actual work on the thing was done by a talented associate who never invented anything on his own. But there were production delays, and he didn't have an exhibit. Later, in 1895, Edison bought another man's projector for a pittance and put his name on it, projected his movies onto a screen, and became the world's first film producer. But the greatest and least-recognized event of the exposition — because it was not yet a consumer product, and the other novelties were — was the triumph of alternating current. Thanks to the example of the Swiss and Germans in Frankfurt and the rotating converter that allowed the use of Tesla's technology on Westinghouse's machines, Westinghouse had won the Chicago contract to light up the fairgrounds.

It had been, as Wellington said at Waterloo, a close-run thing. Under the Chicago contract, Westinghouse was required to install the world's largest powerhouse, twelve single-phase dynamos of a thousand horsepower each. He had underbid, for the Panic of 1893 was raging in the streets. He was overextended, losing money, and it was difficult to see how he would hold on. The contract also required him to provide and illuminate 250,000 light bulbs. Here Edison had his one true moment of revenge. His patent on the electric light had been upheld in court. Everyone else's light bulb was now illegal. And he wouldn't sell a single bulb to Westinghouse. But Westinghouse was no ordinary adversary. A swift search of the literature revealed that the Edison patent covered a globe "covered entirely with glass." The literature also revealed that a valid patent still existed for the carbon-filament Sawyer-Man light bulb, invented by the drunken and difficult William Sawyer, with the backing of Manhattan attorney Albon Man. The Sawyer-Man bulb,

Westinghouse was quick to see, was a "stopper bulb," closed at the base by a plug of ground glass. It was not covered entirely with glass. Hemorrhaging money by the second, he quickly obtained the rights to the rival bulb, set up a manufacturing line, and assigned a team — including Edison's former chief chemist, Reginald Aubrey Fessenden — to make some improvements in the thing. Even so, at ruinous expense, the Sawyer-Man bulbs would require frequent replacement. But the thing — lighting up the fairgrounds — could be done without infringing Edison's patent.

Admission to the fair was fifty cents, and before the exposition closed in the fall, it was visited by the equivalent of a quarter of all the people in the country. On opening day, a miracle occurred. President Cleveland presided over the ceremonies. Although the White House had been laid with electricity for years, no president had ever been allowed to touch a switch. Now, with no ill effects, President Cleveland extended his finger and pressed a small gold-and-ivory key. The Electric Fountain sprang to life, and the American flag automatically rose to the top of its pole. Alternating current was performing work. At night the fountain was illuminated by ever-changing colored lights and the great buildings were outlined by 120,000 bulbs. There was an electric railway and two moving sidewalks, one going three miles an hour, the other six. These, together with a Tower of Light and electric-powered gondolas in the lagoons, were the General Electric Company's contribution. And at the end of an evening, another wonder: in unison, all the lights blinked three times to signal the closing hour. George Westinghouse had proved his case. Universal power was an accomplished fact. In the Westinghouse exhibit, Tesla, almost seven feet tall in cork-soled shoes, passed 250,000 volts through his body and astounded the multitudes with the party tricks from his lab.

But not everything was entertainment. Meeting at the exposition, a congress of electrical scientists from around the world formalized the nomenclature of electrical measurement: the amp, the watt, the volt, the farad, the henry. They were, in order, the basic unit of electrical current, the rate at which electricity is used or wasted, the velocity of electricity, the measure of something's ability to store electricity, and the unit of inductance. The tesla, measuring magnetic fields, would be added later.

Visiting the exposition, the New England aristocrat Henry Adams saw in electricity the death knell of his world and his class. "Here opened another totally new education," he wrote in *The Education of Henry Adams*, in a chapter entitled "The Dynamo and the Virgin,"

> which promised to be by far the most hazardous of all. The knife-edge along which he must crawl, like Sir Lancelot in the twelfth century, divided two kingdoms of force which had nothing in common but attraction. They were as different as a magnet is from gravitation, supposing one knew what a magnet was, or gravitation, or love. The force of the Virgin was still felt at Lourdes, and seemed to be as potent as X-rays; but in America neither Venus nor Virgin ever had value as force — at most as sentiment. No American had ever been truly afraid of either.[4]

But electricity struck fear in his heart. Where some saw delight and others saw boundless wealth, Adams saw nothing but sadness. The world would never be the same again.

* * *

There were other changes. Wiring the Chicago exposition, building the dynamos, manufacturing the Sawyer-Man bulbs, and replacing them when they burned out had just about run through all the money at Westinghouse's fledgling electrical operation, and it was in danger of going under. August Belmont, the transportation magnate, agreed to rescue it, but at a price. The Tesla agreement with Westinghouse on royalties for his patents had to go; no company on earth could afford to pay a single man billions of dollars, which was what, astoundingly, Tesla's agreement amounted to. When matters were made plain to Tesla, according to his own account, he gallantly tore up the agreement with a dramatic flourish. He always regarded George Westinghouse as the only great man in his life who didn't betray him. Still, he later made it clear that he didn't regard the $216,000 he received in exchange for his noble deed as enough. He was living in a cheap hotel with his pet pigeons then, forgotten by most except when he spoke to reporters on his birthday and made extravagant claims for inventions that never materialized. With one exception, a novel and revolutionary new turbine he was unable to build and nobody else could ever make work, he never completed another project.

Edison, too, found himself in altered circumstances. Despite the Westinghouse assault and Edison's fantastic notions of corporate management — not keeping books, for example, and paying his bills by weighing them — the Edison enterprises had done well for themselves in the 1880s. With electricity all the rage, it was almost impossible for them not to. Still, the structure of the enterprise, with the investors owning some things and Edison owning others, and the investors licensing Edison's own technology to the man who had invented it, was sadly in need of revision and rational structure. Moreover, although Edison had built a workable electric train at Menlo Park and had run it on tracks in a lot near the Tabernacle, he had badly missed the boat on electric transit. Encouraged by the railroad tycoon Henry Villard, he'd tried to build an electric locomotive, but like Thomas Davenport, he tried to build it too soon. The future of electric transit lay in street railways, and Thomson-Houston dominated that industry. Villard had an advantage not shared by many men. He was one of the few people Edison would listen to, and he took upon himself the task of bringing order out of the organizational chaos while filling a gap in the product line by buying the Sprague Electric Railway and Motor Company. The resulting corporation, capitalized at $2 million, consisted of the illuminating company, three of Edison's manufacturing companies, and Sprague. It was called the Edison General Electric Company, and Villard was its head. Edison, for his part, was glad enough to get out of manufacturing and back to inventing in his splendid new lab, and he seemed to have very little idea of what Villard had really done. In fact, whether or not Edison actually understood what was happening, Villard had secretly sold Edison General Electric down the river to the Germans. He was a naturalized citizen of German extraction, and it was to Germany that he had fled to regroup and resupply himself with money when he lost control of his railroads in the mid-eighties. He also evolved a plan: a German-controlled electrical cartel based on the Edison patents and the Edison technology. Sixty-two percent of the Edison General Electric Company's stock was owned by the Deutsche Bank.

Morgan had other ideas. He didn't like foreigners except the French. He detested the cut-throat competition of the time and wanted the wealth of the country placed in the hands of men like himself (or, barring that, in the hands of men he could control). And although his opinion of Edison's antics during the Westinghouse struggle can only be surmised, they can hardly have been friendly. He was beginning the campaign of corporate consolidation that would reach its apogee with the creation of United States Steel, and it was clear that his electrical company had, thanks to an inflexible egoist, deployed the wrong kind of electrical system. Westinghouse was unavailable. The country's other powerful electric company was Thomson-Houston. It owned the needed AC technology, and in the person of Elihu Thomson it had an inventive genius who could replace Edison. Accordingly, Morgan laid plans to acquire the company. Word of these plans reached the ears of Charles Coffin. He decided that he and Mr. Morgan had better have a little talk.

In the end, after attracting Morgan's attention, Thomson-Houston bought Edison General Electric, and Charles Coffin was the head of the new company. Mr. Morgan had been impressed by Coffin's logic. Not only did Thomson-Houston possess the necessary AC technology, but it was the richer company, much better organized, and very well run by Charles Coffin. The combined entity was named General Electric, and Charles Coffin began the tradition of superb management that characterized the company for the next century. Edison was outraged. "Tate," he told his secretary, "if you want to know anything about electricity, go out to the galvanometer room and ask Kennelly," referring to Arthur Kennelly, who had replaced Batchelor as right-hand man; later, with the Englishman Oliver Heaviside, Kennelly would discover the ionosphere. "He knows far more about it than I do. I've come to the conclusion that I never did know anything about it. I'm going to do something now so different and so much bigger than anything I've ever done before, people will forget that my name ever was connected with anything electrical." And he vowed to never have anything to do with electricity again. It was a vow he didn't keep. To the end of his very long life, he remained a supplier to the General Electric Company.

He mined his iron in New Jersey and lost one of his shirts. He went into the cement business, invented a concrete house, tried to invent a concrete boat, and lost another one. In the motion picture business, he formed the Edison Trust and dominated the industry. It wasn't entirely about money. It was also about the dogmatic way Edison perceived the motion picture audience. He believed that an audience wouldn't sit in a theater for more than twenty minutes, and the Edison Trust wouldn't allow anyone to make a movie more than two reels long. To escape him, the finest minds in movies fled to California, following a trail blazed by Cecil B. DeMille. Abandoned by his rebellious followers, unable to impose his will, Edison left the movie business in 1910, but not before he made the first filmed *Frankenstein*.

Following a few overdramatized all-night sessions to finally improve his phonograph after the West Orange lab opened in 1888, he reentered the phonograph business with his perfected machine. He hired second-rate singers, because he believed that people weren't interested in listening to a star. When he learned that the customers were amusing themselves by playing the cylinders at variable speeds, he installed a governor that made it impossible. He liked straightforward music; he detested the complexity of opera and the rich piano of the Romantic movement. Although he was increasingly deaf, he insisted on interviewing the artists himself; he listened to a piano by biting it, so that the sound wave of the music reached his brain through the vibrations in his bones, a sight that once reduced the founder of the Montessori movement to tears of pity. Stubbornly, he stayed with cylinders, which he had invented, while others moved to the flat disks that became the industry standard. He left the phonograph business in 1929.

Believing that the internal combustion engine was an interim technology, he developed the alkaline storage battery, testing his products by dropping them out of windows. Elsewhere, electrical science advanced without him. He thought the radio was a passing fad. Like Tesla, he was a man of the nineteenth century, and he never got over it. He'd been gone from his company for a year, and it no longer bore his name, when the city fathers of Buffalo, New York, decided to enter the twentieth.

In the last decades of the century, Buffalo was an American go-getting town. With silos fed by the rich harvests of the Genesee Valley, it was a major grain center. It was second only to Pittsburgh as an iron and steel center, second only to Chicago as a rail hub. In 1886 it bought one of George Westinghouse's first alternating power systems, and in 1890, as Kemmler walked his last mile, the city's great project was to harness the immense power of Niagara Falls. The previous year, Morgan, John Jacob Astor IV, and other capitalists had formed the Cataract Construction Company to explore possible forms of large-scale exploitation using advanced technology. Westinghouse, at first, thought a compressed-air system would do the trick, and four proposals were submitted to do just that. But the chairman of the jury that would determine the method was the former William Thomson, now Lord Kelvin. In recent years, Kelvin's opinions had grown somewhat strange. He had not only tried to refute Darwin, but as president of the Royal Society he had declared heavier-than-air flight impossible. But he still knew a great deal about electricity. In Britain, with its crippling laws restricting the spread of electricity, the study of the subject had become an academic exercise, but it was an extremely good one.

Originally, Kelvin favored direct current. So did Edison, who had not yet been separated from his company by Morgan. But direct-current power plants — there would have to be a large number of them, given the power of Niagara — would have to be clustered near the falls. Then a new industrial area would have to be built, also near the falls, and the new industries could not be powerful ones; for direct current was reliable and cheap, but it was weak. DC-powered clothing factories were possible, for the electricity could run the sewing machines; many other kinds of factories were ruled out. But the question was moot. In 1890, when the Germans ran their 108-mile electrical line, direct current became history. Alternating current could run just about any kind of factory, and because it could be transmitted for long distances, it could be brought the twenty miles to Buffalo itself. The new rotary converters could hook it into the existing power grid and the electrical lines that ran the city's trolley-powered mass transit system, the high voltages would bring powerful new industries, and Westinghouse was the man of the hour. Kelvin's jury awarded him the first major hydroelectric contract in the history of the world. With it, Westinghouse would build the world's first universal electric service. It would be alternating current all the way.

It was not quite the victory it seemed. General Electric, moving rapidly into AC under Coffin and Elihu Thomson, built the power lines and the next set of dynamos, and between General Electric and Westinghouse, peace was declared. It came not a moment too soon, because GE and Westinghouse were in a position to immobilize each other, not merely in Buffalo, but in the entire country. Westinghouse controlled the AC patents. GE controlled the patents for the electric light. Westinghouse had turned a blind eye when GE — with the brilliant German dwarf and engineer Charles Proteus Steinmetz added to its growing scientific staff — pretended that it had developed a new form of polyphase, but he could stop the company in its tracks anytime he chose. But GE could stop him with equal ease. Now that Westinghouse's Sawyer-Man light bulb patent had been overthrown in court, GE could throw Westinghouse out of the electric light business anytime it chose. The situation was absurd. Reasonable men were now at the helm in both companies, and they solved their problem by pooling their patents. It was time to let the dead bury the dead.

And it was time to move on. To Westinghouse, Elihu Thomson, and Steinmetz, the shape of things to come was obvious. Niagara was a unique resource, and not every city had access to hydropower. Other means of generating and transporting electricity would have to be devised. Dynamos had grown to the point where the techno-logical limit of the exterior steam engine had been reached. With an exterior steam engine, the engine caused an overhead drive shaft to rotate by means of a belt, and the machines on a factory floor were also powered by belts leading down from the shaft. Friction and the tendency of the belt system's components to fly apart when driven beyond a certain speed meant that the work done by an ex-terior engine had hit a wall, and no increase in electrical power could be expected from that source. The next new thing would be the steam turbine and limitless amounts of electrical power. Al-though they were no longer young, it was still possible to do things no one had ever done before.

Meanwhile Buffalo, with its immense new amounts of electric-ity, became the aluminum-smelting capital of the nation, and it re-mained so for years. Buffalo, like Paris, became the City of Light. It was immensely proud of itself, and to celebrate, it did what a city did in those days: it held an exhibition and invited President McKinley. He was shot by an assassin at the railroad station.

Tesla, whose patents had done so much to make the Niagara project possible, did not attend the exhibition. Although his statue, donated by a country, Yugoslavia, that no longer exists, graces Goat Island above the falls to this day, he didn't attend the opening of the first powerhouse, either. He was much too busy trying to pry a large sum of money out of John Jacob Astor IV. Tesla had a new plan. It would astonish the world.

CHAPTER NINE

MARCONI AND THE LETTER S

Time will tell, Mr. Marconi.

— Tesla

Guglielmo Marconi saved the Eiffel Tower. In 1909 the Paris landmark, having long outlived its apparent usefulness, was scheduled for demolition when it was discovered that the country's master radio antenna was located at its peak. Radio had become so essential to the life of a modern nation that the demolition project was put on temporary hold until the antenna could be relocated. The project remains on temporary hold to this day.

But Marconi did not invent radio. He was an amateur engineer who assembled it from the inventions of others, and he never laid claim to being anything more. He was using the French physicist Edouard Branly's coherer (as improved by the English physicist Oliver Lodge, it was a tiny, crude device that conducted electricity with glass-enclosed metal filings) to detect the radio wave, and when he claimed he had improved it further, he had actually re-placed it with a coherer invented by Jagadis Chandra Bose, the greatest Indian scientist of his day. He had been anticipated by Edison, who sold Marconi his own rather useless radio patent in

1903. His tuning apparatus, which allowed him to separate the signals, had been invented by Nikola Tesla, Oliver Lodge (again), and the curiously named American researcher John Stone Stone. Most of these contributions he acknowledged. Unlike Edison, he did not tell fibs, and unlike Tesla, he did not neglect to inform his backers of what he was doing. There was, however, a signal omission from his catalog of borrowings.

"Yes," Tesla said happily, "[Marconi's] using seventeen of my patents." And Marconi had no right to use Tesla's patents.

There was ample reason for Tesla's equanimity on the subject, however. He had worked in radio all through the 1890s and he held all the fundamental patents, but as the turn of the century approached, Tesla was no longer fooling around with radio. He had decided that radio was beneath him. Now, with Morgan's money and certain new groundbreaking discoveries, he was preparing to destroy Marconi as he had already destroyed Edison. Once again, Tesla was firmly convinced, he had developed by far the superior system. His new wireless communications device, when it came, was going to be something tremendous.

* * *

In Russia another physicist, Alexander Popov, wanted to know what all the fuss was about. So Marconi was using a Branly coherer. So what? Popov's own radio apparatus used a Branly coherer. The young Italian Marconi had deployed a spark-gap transmitter? So had Popov. Marconi had transmitted his signal over a distance? But Popov had already done that. Popov was mildly annoyed, but not at Marconi. Like Henry, Faraday, and Dr. Chandra Bose, Popov made his discoveries for the benefit of the world, but he had failed to give the world adequate warning of what he had done. Russia was remote from the scientific centers of Europe, Popov did not publish his most dramatic findings, and outside of his immediate circle, almost no one knew what he had done.

Marconi, following Morse's precedent, made his discoveries for the benefit of himself.

* * *

In 1896, when the twenty-two-year-old Marconi and his Scotch-Irish mother, the former Annie Jameson, arrived in England with their primitive device for the wireless transmission of electrical impulses, wireless transmission was already sixty years old.

Working in India in 1839, an East India Company surgeon named William O'Shaughnessy had independently invented the telegraph, set it up, and made it work. He completed his circuit with two and a half miles of the Hooghly River, using the conductive powers of water. It appears to have been the world's first wireless transmission.

Morse did much the same thing in 1842, when he sent a signal through the waters of the Chesapeake and Potomac Canal in Washington, D.C. In 1838 Steinheil — who discovered the earth return that was incorporated into Morse's telegraph by Vail and Cornell with such good results — postulated that ground transmissions were possible and actually sent a signal fifty feet. In 1882 Alexander Graham Bell used the waters of the Potomac River to send a telephone message for a mile and a half. These were but a few of the precedents.

Nor were airborne electrical transmissions unknown. In the parlance of the time, they were called "electricity at a distance," and they usually manifested themselves as a spark or a measurable current in some conductive material. Joseph Henry observed electricity at a distance in the 1830s, Faraday probably did so, too, and Edison rediscovered the effect in 1875. In Henry's last major experiment before he answered the call of duty, moved to Washington, took up the reins at the Smithsonian, and destroyed himself as a practicing scientist, he had also discovered — using the metal roof of his house as a receiver — that atmospheric electricity could be used to magnetize needles in his basement. It was, of course, the same atmospheric electricity that Franklin had observed a hundred years earlier.

Some had also speculated that airborne electricity could be made to perform work. The English inventor Edward Davy had postulated that church bells could be made to ring at a distance of a mile, and he may actually have done so. Using kites and the conductive properties of cloud cover, Dr. Mahlon Loomis, a dentist, had succeeded in sending electrical impulses from the Catoctin Ridge to the Bear's Den Mountain in Virginia between 1866 and 1873. Loomis deceived himself into thinking that he could broadcast from the Rockies to Switzerland for the purpose, among other things, of "converting the heathen." He tried and failed to obtain a congressional subsidy, and died a bitter man in a little West Virginia town, where he spent the last years of his life running a tiny broadcasting studio. Sometime around 1885, Amos Dolbear, a physics professor at Tufts, had sent a message for a mile, probably by induction, the sort of electrical activity caused by rotating a bar of iron in a magnetic field, and claimed that he had sent one thirteen miles, which almost certainly meant that he was using radio waves. Edison, also using induction fields, had sent a telegraph signal in 1887 from a moving train car to the telegraph line that ran parallel to the track, and passive induction carried it along the wire. In England, William Preece, chief engineer of the Royal Mail, which in addition to delivering the nation's letters and periodicals now owned and ran the British telegraph and telephone systems, had transmitted an induction signal for six miles. And the Slovakian-American priest Father Josef Murgas had built a wireless telegraph system of some sort in Wilkes-Barre, Pennsylvania, although he didn't get around to demonstrating it until 1907.

Even more promising developments were afoot. The Scottish mathematician James Clerk Maxwell had mathematically surmised the existence of electromagnetic radio waves in 1867. On that level, things were stranger than anyone had surmised. Maxwell postulated that radio, like light, traveled in waves. From this, it could be concluded that the wave could be shaped. There was a very intriguing possibility in this. Sound — such as the sound of the human voice — also travels in waves. If a radio wave could be formed into the shape of a sound wave, it should be possible to broadcast the human voice without wires. But first there had to be a real-world demonstration of Maxwell's mathematical speculations, and in 1886, in a series of remarkably elegant experiments, the German physicist Heinrich Hertz had demonstrated them in the lab.

He set up two metal rods end-to-end with a small gap between them. In this gap was where the radio wave appeared in the form of a spark that moved back and forth — that is, it oscillated, and this oscillation was the sure sign of a radio wave. Across the room, he set up a loop of wire pierced by another small gap. When the radio wave appeared there, oscillating, he had performed a broadcast. And the lab, Hertz believed, was where radio waves would stay. Their range appeared to be about three yards. "What," Hertz memorably asked, "is really the purpose of the whole nonsense?"

Sometime around 1895, Alexander Popov chose to find out.

Years of Soviet-American rivalry have hopelessly obscured the fact that, from the 1890s to the First World War, the Imperial Russian Torpedo School in Kronstadt was, along with the Royal Institution and Edison's Menlo Park, one of the most advanced electrical research facilities in the world. Much of the early work on electronic television was done there, where Alexander Popov was an instructor.

Using Maxwell's equations, Hertz's papers, and a Branly coherer, Popov set out to create a detector of distant thunderstorms. And in many ways, the early story of the wireless is the story of that very coherer. In the 1850s the British telegraph engineer S. A. Varley was plagued with lightning; it kept striking his wires and shutting down his systems. He discovered, however, that the filings of certain metals, such as iron, had curious properties. In the absence of atmospheric electricity, they were inert and nearly impervious to an electrical current. In the presence of atmospheric electricity, their resistance fell sharply; the metals suddenly cohered. That is, they stuck together. In other words, the iron filings had detected the electricity, and they could be used as a switch, because when they stuck together in a small container, they behaved like a single piece of conductive metal. Here was the solution to Varley's lightning problem. He attached the little container of iron filings to his telegraph wires and connected it to the ground with another conductive wire.

When the telegraph wire was struck by lightning, the iron filings cohered and carried the surge of electricity to the ground wire, which conveyed the lightning harmlessly to the ground. Branley, Popov, and Bose used this switch in their experiments, and in 1894 Sir Oliver Lodge, the spiritually inclined British physicist who invented the automobile spark plug, used (and named) the Branly coherer to detect Hertz's radio waves — which were then called Hertzian waves — at a distance of 160 yards. Although Lodge didn't realize it until later, when he began to patent everything he could think of, he had invented the radio. To be sure, he was not the first to do so. Tesla was also inventing the radio, and Tesla knew what he was doing. Lodge made one more contribution. He equipped the coherer with a "tapper," a malletlike device that struck the tube, de-cohered the iron shavings, and prepared them to receive the next signal. It was with this tiny device that Popov set out to detect the approach of a thunderstorm.

Anyone long enough in the tooth to remember old-fashioned vacuum tube AM radios will realize at once that a thunderstorm-detector and a radio are identical, because at the approach of a thunderstorm, an AM radio begins to give out bursts of static. In the Branly-Popov system, the thunderstorm manifested itself as it did for Hertz, as a spark in a gap between two wires. It was not long before Popov realized that if he could detect thunderstorms, he could also send messages in Morse code. The duration of the spark would indicate whether it was a dot or a dash. To his Branly-Popov apparatus, Popov added a Hertz transmitter. He had built a wireless telegraph. On May 7, 1895, many months before Marconi had broadcast for more than a mile, he had sent a thirty-mile signal from a Russian warship to his lab in St. Petersburg. The navy kept the experiment quiet.

Next, on March 24, 1896, just as Marconi was arriving in England accompanied by much fanfare, Popov demonstrated his device for his colleagues. As President F. F. Petrushevsky of the Russian Physical and Chemical Society stood at a blackboard with a piece of chalk and a dictionary of Morse code, one of Popov's students in a building eight hundred feet away tapped out a message. Consulting his dictionary, Dr. Petrushevsky slowly wrote the words HEINRICH HERTZ. It was Popov's tribute to the borderless world of science. It was also the first aerial — as distinct from water- or earthborne — wireless message in history.

But although Popov's wireless sets were, none too swiftly, installed by the Imperial Navy, Popov himself was uninterested in exploiting his invention for commercial purposes. Instead, he turned to the study of the German scientist Wilhelm Conrad Roentgen's new X-rays, joining Edison and Tesla in perfecting this latest marvel of the age. When Marconi, covered with glory, visited Russia, Popov was his amiable host, and the young businessman and the middle-aged scientist sat down to have a good laugh about impractical dreamers such as themselves. When Marconi's impending marriage to Beatrice O'Brien, daughter of the fourteenth baron Inchiquin, was announced in 1905, Popov sent him a fur coat and a silver samovar.

Popov died in that same year, his good liberal heart broken by the reactionary tsarist government.

Marconi became a fascist.

* * *

Marconi was an aristocrat and a philanderer who happened to know a lot about electricity. Around the Villa Grifone in Bologna, he was known by his English nickname Billy unless his stern Italian father happened to be in the immediate vicinity. And his mother, a member of the Scotch-Irish whiskey-distilling family of Jameson, saw to it that he received his training from the finest scientific minds in the land. It was just as well that she did. He had failed to get into the naval academy. He had failed to get into the University of Bologna. In his early twenties he was interested in sailboats, fishing, women, and electricity. His only visible accomplishment was a pitch-perfect command of British English. He had never done a single constructive thing in his entire life, and according to his angry and despairing father, he never would. In his father's view, the young man's obsession with electricity was a meaningless sideshow.

And so, with his mother's connivance, young Marconi locked himself in the mansion's third floor, took his meals off trays, and conducted his wireless researches in secret.

He was advised in physics by, among others, Professor Augusto Righi, a family friend and nearby neighbor who carried on Hertz's work and taught him that the best results were obtained if he coated his apparatus with Vaseline. The Marconi apparatus in question was the same apparatus used by Heinrich Hertz. In 1895, supposedly inspired by Hertz's obituary, written by Righi himself, Marconi had succeeded in duplicating Hertz's experiment, projecting electromagnetic waves across a room, where as usual they manifested themselves between the points — balls, actually — of a detector as a tiny spark in a gap. Then he taught himself to ring a disconnected electric doorbell at a distance of thirty feet, again with the invisible Hertzian electromagnetic waves. Slowly, he was working in the direction of sending and receiving a message. He couldn't believe that no one else seemed to have thought of it; perhaps there was some crucial gap in his reading, or perhaps the Americans, who seemed to be doing the most worthwhile work, had done it and kept the secret. And an American had. Starting in 1893, Tesla had begun to file certain fundamental patents in radio. They were hardly secret, and Marconi was soon making use of them. All his life, he was candidly aware of his own limitations. He was not a scientist, he said. He was an amateur, and in the mid-1890s it beggared his imagination that he and he alone, the failed wastrel son of an Italian country gentleman, had conceived the idea of transmitting intelligence by Hertzian waves. Then he stumbled on the Branly coherer. The Branly-Lodge coherer became Marconi's signal detector, just as it had become Popov's. With the addition of the coherer, Marconi's prototype was nearly complete.

Radio, like visible light and water in a body, travels in waves. Water, as O'Shaughnessy and Morse had demonstrated, carries electromagnetic transmissions, and light waves, if they are long enough, can be heard by a radio as well as seen by the eye, because

they have become radio waves. And for transmitting radio waves —
Hertzian waves — a whole new vocabulary had to be invented. Radio waves are the longest wavelengths in the electromagnetic spectrum, waves that can be as short as a meter or as long as a mile
(although there is no upper limit). The word "spectrum" again leads
us back to the behavior of light. When light is broken into its components by a prism, the resulting colors are said to make up a spectrum. A similar thing happens with radio waves. When a radio
wave travels through the air of the earth or through interstellar
space, it travels through a position in the radio spectrum, a sort of
invisible skyscraper with many floors, where a radio wave can occupy only one floor at a time (and there are many radio waves).
Which floor — which part of the spectrum — is determined by the
length of the wave. Although the reality of the situation is somewhat more complicated, as is often the case, industry parlance for a
century has spoken of going "up the spectrum," speaking of shorter
and shorter waves — designated by rising frequency number — the
higher up in the metaphorical skyscraper one goes. In the lower,
easiest-to-use part of the spectrum, the waves are long, and these
long waves were what Marconi was using. As the waves go higher in
the skyscraper, they grow steadily shorter, and the shorter they become, the harder it becomes to use them for the transmission of
sounds. Finding which part of the skyscraper — that is, which part
of the broadcast spectrum — a particular radio wave occupies is
called *tuning*.

A person with a receiver has to determine how long the radio
wave is because it can be found only in its particular part of the
spectrum. This is what happens when you turn the dial on your car
radio. You are finding the wave, whose location is determined by its
length, and when you've found it, you can listen to the broadcast
carried by that particular wave. The radio waves, like light (and
electricity) traveled at 186,000 miles per second, the upper speed
limit of the known universe: nothing had ever been observed
that went faster. But young Marconi was far from exploring the

mysteries of radio tuning or pondering the speed of the waves. At this point in his young career, he was simply trying to find how far a radio wave could go.

The coherer enabled him to extend his range, first down the stairs, then out to the terrace, and finally out to the garden. He confined his experiments to the morning and the evening; at midday he was likely to be underfoot of the servants or, worse, to run afoul of his father. Early on, he learned that his battery power supply was less important than the sensitivity of his receiver, and he accidentally stumbled on the importance of the earth return and the antenna, neither of which had occurred to Popov. The symbiosis of the trickle power of a battery and the sensitivity of a receiver had been known since the days of Morse and Vail and had been dramatically demonstrated by William Thomson in the Atlantic cable, although Marconi had to discover it for himself. The principle of Steinheil's earth return had been known for sixty years. The antenna was something new. With an antenna, a bit of conductive metal held aloft above his Hertzian apparatus, he could both send a signal and receive one.

"By chance I held one of the metal slabs [used to increase the wavelength] at a considerable height above the ground," Marconi recalled, "and set the other on the earth. With this arrangement the signals became so strong that they permitted me to increase the sending distance to a kilometer." The slab of metal he held aloft above the ground was a capacitor, a metal sandwich that also served as a temporary electrical storage device. Soon he discovered that, to make an antenna, it was only necessary to hold aloft a commonplace wire. Meanwhile, in America, Tesla was able to send his signal for thirty miles. But nobody was paying much attention to Tesla's radio experiments, because he kept them to himself and didn't publish his findings.

Prevailing scientific wisdom said that an electromagnetic wave would be blocked by a building, a tree, or a hill. Marconi, employing his brother and the gardener as his assistants, proved otherwise. When they could see each other and a signal was detected, a flag was waved. When they were out of sight, a gun was fired. Marconi believed that his waves could penetrate obstacles. It wasn't until much later that he and everybody else discovered that they followed the contours of the landscape.

The Italian government wasn't interested in his system; the telegraph, the government told Marconi, was adequate for its purposes. But his Irish mother was not exactly a colleen of the people. She had been born in Daphne Castle, County Wexford, and her prosperous family had more-than-adequate contacts in the British Isles, where she now conveyed her more-than-willing son. The British customs officials were suspicious. No matter how fluently he spoke the language, young Marconi was clearly an Italian. Italy was the land of the anarchists, and Marconi's device, with its mysterious wires, closely resembled an infernal machine, the era's somewhat colorful term for a bomb. Fortunately, Marconi also came equipped with a Jameson from Daphne Castle, who was not about to tolerate the tergiversations of some uniformed jack-in-office. One imagines many reddened ears as mother and son swept in triumph from the customhouse.

Through family connections, Marconi was soon in touch with the Royal Mail's William Preece, the sixty-year-old Welsh telegraph engineer who is largely remembered for two statements of staggering fatuity. "The Americans have need of the telephone, but we do not," he said. "We have plenty of messenger boys." He also characterized Edison's electric light as a "completely idiotic idea." With the nationalized British telephone system added to the previously nationalized British telegraph system and placed under the control of Preece's post office, he was now using this sort of uninspired thinking to do everything in his power to retard the development of his country's communications technology, although he had become deeply interested in the phenomenon of crosstalk — the interruption of one telephone line by another. To the telephone user, crosstalk was an annoying accident, but Preece was experimenting with it as a possible method for creating a wireless telephone system. He wasn't getting very far. Using induction fields, he found that he could send a message for a distance of five miles, but only if the wires were parallel and equal in length. Obviously this sort of thing wasn't going to create a nationwide wireless system anytime soon, and young Marconi's Hertzian apparatus called to him, because it seemed to offer the wireless promise that crosstalk didn't. Marconi's radio didn't require wires in a special configuration, it seemed to broadcast from anywhere, and experiments soon suggested that its range was indefinitely expandable. Preece placed the considerable resources of the Royal Mail at Marconi's disposal, and he lent him the services of the man who would become Marconi's faithful sidekick for the rest of his life, William Kemp. With Preece's patronage, his Hertzian apparatus, his Branly coherer, and his new antennas, Marconi sent a signal across Salisbury Plain, bridged the Bristol Channel, reached the coast of France, and encountered his first schizophrenic, a young Frenchman who insisted — at gunpoint — that the wireless waves were disturbing his nervous system. In July 1898 the Dublin *Daily Express* became the first newspaper in history to obtain a dispatch — describing the outcome of the Kingston regatta — by wireless telegraph. "Is it an Irish characteristic," the paper asked,

> or is it the common impulse of human nature, that when we find ourselves in command of a great force, by means of which stupendous results can be produced for the benefit of mankind, our first desire is to play tricks with it? No sooner were we alive to the extraordinary fact that it was possible, without connecting wires, to communicate with a station which was miles away and quite invisible to us, than we began to send silly messages, such as to request the man in charge of Kingston station to be sure to keep sober and not to take too many whiskey-and-sodas.[1]

In August wireless was installed on the royal yacht, and Queen Victoria soon had news that Albert Edward, the Prince of Wales, had been injured in a sailing accident. For a few seriously strange weeks, the sole topic of conversation over the greatest invention of the age was the state of Bertie's knee. The following March, wireless saved its first lives, when the East Goodwin lightship was rammed in the fog and, thanks to wireless communications, its crew was saved.

Tesla watched with amusement from afar.

* * *

It has never been very clear what Tesla was doing in Colorado Springs during the years 1899–1900. He had gone to the remote location because he believed he had almost destroyed New York City with a man-made earthquake, or so he claimed. The dangers of urban experimenting had grown too great, and a friend who was a director of Colorado Springs' electric company offered unlimited free power. Tesla built a tower on the outskirts of town. His new goal was the wireless broadcast of electricity itself. Every experimenter knew that you could wirelessly send small amounts of power, usually inadvertently, for small distances. Tesla knew that he could send a radio signal for thirty miles. Using the nickel-iron core of the earth as his conducting medium, he now planned to cheaply send massive amounts of electricity and floods of messages to every location on the planet. He would become rich, always a paramount Tesla goal, and he would be the greatest benefactor of mankind who ever lived.

He had always been a showman. In his New York laboratory, demonstrating the harmlessness of alternating current, he had manipulated glowing orbs of pure electrical energy with his bare hands, sometimes passing them to his friend Mark Twain. It is still a little unclear how he did that. In Chicago at the World's Columbian Exposition, dressed in faultless evening clothes, he had nightly turned himself into a human conductor, linking Westinghouse's huge turbines to a motor by passing the electricity through his body. In that same year of 1893, well before Lodge and Marconi, he had broadcast Hertzian waves across the stage of a St. Louis lecture hall, and in Madison Square Garden in 1897 he had demonstrated a radio-controlled boat that seemed to have a mind of its own. "I was, of course, controlling it," he later assured his readers. He understood how to tune his signal — Lodge and Marconi had only a rudimentary knowledge of that — and he had taken out the relevant patents. Even where there was only one radio signal in the air, you needed tuning to know where it was. If you were sending more than one radio signal to a single object — to operate, say, the radio-controlled motor but also to blink its lights — tuning was the only way to do it. And if there was more than one radio signal in the earth, tuning would keep them separate. Tuning made possible a multitude of radio stations. Tuning was made both possible and essential by another strange conductive feature of the atmosphere, because it carried the radio spectrum, and the radio spectrum, as previously noted, was divided into a number of invisible highways.

If you sent an electrical signal into the water, it went in all directions, but recovering it was simple. You stuck a wire in the water and listened to the wire. The same was true of a signal sent through the earth. The atmosphere was different. True, a radio signal sent through the air went in all directions, which is why a message sent from Berlin to Tokyo could be overheard in London and New York, but there the similarity to earth and water ended. A radio signal sent through the air went through a specific portion of the air and could not be heard in other portions of the air. There were many pathways, they were mutually exclusive, and we now call them frequencies. If you wish to send a message, you will send it on a specific frequency. Only a receiver tuned to that frequency can hear it. Tesla understood this principle. Indeed, he probably knew more about it than anybody on earth. Given his knowledge and his gifts, he could probably have invented the form of radio toward which Marconi was heading. But he had decided not to. Hertzian radio waves, he had decided, were needlessly wasteful, because they went in all directions and did only one thing. They did not, for example, transmit usable household electricity, they could not run the world's telephone systems, and they could not send printed messages. Tesla believed he had conceived of a system that would do just those things.

Now, in Colorado Springs with $100,000 of Colonel Astor's money, he commissioned a local carpenter to build him a lab attached to a tower. The tower was not for the broadcast of radio waves. It was for the production of lightning. Once the lab and the tower were completed and vast amounts of outrageously expensive equipment had arrived from the East, the nocturnal grounds around them glowed with St. Elmo's fire and the municipal fire plugs created an arc whenever a screwdriver was passed close by. Tesla was pumping electricity directly into the earth, and these were the first results. Colonel Astor, that most henpecked of men, believed he was investing his cash in Tesla's revolutionary filamentless, cold lighting tubes — the forerunners of fluorescent lighting — and the mechanical and electromagnetic oscillators Tesla had designed. In Tesla's somewhat misleading terminology, a mechanical oscillator was a self-contained steam turbine, an electrical generator that contained its power plant rather than running on an external steam engine that transmitted its power by means of belts and shafts. The steam turbine, in Elihu Thomson's view, would unlock the next generation of riches — vast riches, because a steam turbine could generate much more electricity and do it very cheaply. Colonel Astor thought he was buying a big piece of the future. As for Tesla's electromagnetic oscillators, they enabled him, in theory, to detect a radio wave; without oscillators, a modern radio set cannot function. A later generation of Tesla's electromagnetic oscillators would be called vacuum tubes, and an even later generation would be called transistors. They would prove to be a bonanza for anyone who invested in them.

Colonel Astor, contemplating all these wonderful inventions, justifiably looked forward to making a bundle from new consumer products, but Tesla, who never understood why Colonel Astor wouldn't give him another cent, was having none of it. Building on the magnetic and electrical discoveries of the first Queen Elizabeth's physician, William Gilbert, he was attempting to use the planet itself to propagate an electrical signal of enormous strength. His plan was a model of elegant simplicity. With his vast electrical signal, Tesla would water the deserts, alter the climate, illuminate the sea-lanes at night, and provide everybody in the world with unlimited quantities of free electricity. It had occurred to him, faintly, that Colonel Astor, a major investor in the Niagara power plants, might not regard this as a very good idea.

After embedding his electrodes in the earth, he would fire an electrical impulse into them. At first, it would be limited by the power made available by the local electric company. When it returned — from the opposite side of the world, he believed — he would fire it back again, propagating it through the nickel-iron core of the planet. With a simple device, easily manufactured, consumers could illuminate their houses with limitless electricity. They could also dial any telephone in the world, send faxes (although he didn't call them that) and telegrams, and synchronize their clocks, although he never got very specific about how he would perform that last miracle. At last, Tesla's tower was ready for its greatest test.

Wearing rubber-soled shoes, Tesla stationed himself at a distance as his assistant, similarly attired, stood by to turn the dynamo off and on at Tesla's signal. A six-foot bolt of electricity, as measured by Tesla's eye, flew from the top of the tower. The man-made thunder, it was said, was audible twenty-two miles away. Gradually the lightning bolts increased in strength — proof, Tesla believed, that his plan was working — until one attained a length of 130 feet, a record still unsurpassed largely because no one else can see the point of the exercise. (It is unclear how Tesla measured it. Let's just say that he got a long lightning bolt.) Then the tower and the entire city went dark. Tesla had burned out the power station.

In Colorado Springs, he also believed he had received radio signals from an alien civilization, perhaps on Mars, something that caused the first crack to appear in his scientific reputation. Actually, he had experienced a phenomenon that is now well known. If you listen to the sky, you will hear radio signals; listening to them is the job of a modern radio telescope. He had detected the background radiation of the universe, all those long light waves that got stretched into radio waves, but neither Tesla nor anyone else knew that.

He returned to New York convinced that he was on the track of the most powerful force in the history of the human race. With his claim, he put up a half interest in his patents as collateral and obtained a $150,000 loan from J. P. Morgan (note that Morgan, unlike Astor, required collateral and had specified what it would be), bought a tract of land called Wardenclyffe near Shoreham, Long Island, and began to build his perfected electrical tower.

Topped by a mushroom-shaped dome, it was 187 feet tall. Tesla also burrowed 120 feet into the earth. He did not intend, he pointedly reiterated, to use the radio waves discovered by Hertz. Because they radiated in all directions, Hertzian waves were weak and wasteful. Instead, Tesla proposed to use electrical current as his carrier of telegraphy, voice, and data. How he was going to do this, he did not explain. The Tesla currents, he claimed, would be 90 percent efficient. Lord Kelvin, the former William Thomson of the Atlantic cable and now a skeptic of just about everything, listened to Tesla's explanation and emerged from the conversation a convert.

At Wardenclyffe, he also created vacuum tubes, although he never revealed what they were supposed to do or whether they even worked. It is possible that they did, because Tesla, unlike Edison, was capable of understanding the Edison Effect — the darkening of the inside of a light bulb that, puzzlingly, took place in a vacuum. With his remarkable powers of visualization, he kept few notes, and when he published in the popular press, especially the *Century* magazine, edited by one of his few friends, he described what he had done or intended to do but never revealed the method. He was no longer inclined to look on Marconi with indulgence. The young Italian was infringing his (and now Morgan's) patents. Marconi had also informed the press that he was using the Tesla coil, the immensely powerful device for magnifying the power of electricity, also invented by Elihu Thomson, amid whose awesome lightning flashes Tesla famously had himself photographed — in a double exposure; the actual lightning would have killed him. To defeat Marconi, and without telling Morgan, Tesla had doubled the size of the tower designed by his own friend Stanford White. He had also neglected to tell Morgan that he proposed to distribute free broad-cast power to the world. Morgan was another major investor in the Niagara project, and he was a man with an extremely specific mind. He had told Tesla exactly what his money was to be used for. Tesla was to wirelessly broadcast yachting and horse racing results, to-gether with stock quotations. That was it. That was what his money was for, and nothing else. Morgan didn't know enough about radio to realize that Wardenclyffe wasn't a radio tower, but he knew how to draw up a contract and he had made Tesla sign one. Tesla was now in breach of that contract. And Morgan had told him, again with great specificity, that $150,000 was all the money he would ever receive. The money was now gone.

On one night in 1901, and one night only, the Wardenclyffe tower produced lightning bolts. Then it fell silent, never to revive again. Tesla, unable to believe his ill fortune, pleaded for more funds.

"Are you going to leave me in a hole?!!" he embarrassingly wrote to Morgan. "I have made a thousand powerful enemies on your ac-count, because I have told them I value one of your shoestrings more than all of them." And again, "Have you ever read the book of Job? If you will put my mind in place of his body you will find my suffering accurately described. I have put all the money I could scrape together in this plant. With $50,000 more it is completed, and I have an immortal crown and an immense fortune."[2]

"My dear sir," Morgan wrote, "in reply to your note I regret to say that I should not be willing to advance any further amounts of money."

"Men are like flies to you," Tesla had written. "Please do not write to refuse."

Morgan wrote to refuse. Tesla raced to his tower, climbed it wildly, and desperately embraced it.

It seemed like such a trifling sum, but a Morgan loan, Tesla found, was also a loaded gun. If Morgan was into a deal with Tesla, everybody except the disgruntled Colonel Astor wanted in. If Mor-gan was out, Tesla couldn't find a nickel in the street. Dodging bill collectors, he attempted to continue his experiments and find new investors. None came forward. His unpaid workmen trickled away until only his faithful assistant remained at his side. His reputation in a shambles, he turned the Wardenclyffe tower over to the Waldorf-Astoria Hotel, where he lived and also neglected to pay his bills. The tower would be dynamited for scrap in 1917.

* * *

In Europe, Marconi was on a roll. The Royal Navy bought experimental wireless sets. So, too, did the Italian navy, seeing the error of its ways. Lloyd's, the British insurance consortium that was also in the lighthouse business, bought sets to communicate with its lighthouses. The lighthouses bought sets to communicate with incoming ships. But the new broadcasting stations were expensive to build and expensive to run, and they were in very odd places, usually wastelands on the seacoast. Throughout the century, the onward march of electrical science had been in the direction of smaller, cheaper, and faster devices, but Hertzian broadcasting was of an entirely different order of magnitude. The broadcasting stations were huge. The capacitors, metal induction sandwiches without which a signal could not be generated, were forty feet tall, and Marconi had thousands of them. England still had no national electrical grid, and Marconi's battery was probably the largest ever made. The spark in the spark gap transmitter was "as thick as a blacksmith's wrist," but the device was, basically, an enlargement of the one Hertz had built. To prevent the operator from being incinerated by it, the key was three feet long. The unholy din was "not unlike a Maxim gun." When the station was transmitting, the crew stuffed their ears with cotton. The resulting commotion recalled the old joke about the man who was talking to London. The punch line: "So I see. But what's he doing with that telephone in front of his face?" In other words, a wag pointed out, a Marconi station made so much noise that it could probably be heard across the oceans without the use of radio waves. Meanwhile, the £50,000 Marconi had obtained from London banks was dwindling. To sustain interest — and make the fortune due him — Marconi needed to do the impossible, and he knew to a certainty what impossible thing he needed to do. He needed to broadcast across the Atlantic. Everyone was unanimous in agreeing that he couldn't do it. The science of his time was against him.

The science of his time believed that Hertzian waves were strictly linear — that is, they would proceed in a straight line until the earth's curvature caused them to leave the planet. James Clerk Maxwell had proved that light and electromagnetic waves were analogous; they traveled at the same vast speed, and everyone, until Einstein said otherwise and Oliver Lodge proved him right, knew that light didn't bend. No one knew that the ionosphere existed, the thin upper layer of the atmosphere that lies above the stratosphere. Heaviside and Kennelly hadn't discovered it yet, although Tesla had surmised it, and even if people had known about the ionosphere, nobody would have known what it did. It bounced radio waves back to earth. Marconi knew that something was going on — he was now broadcasting at ranges of hundreds of miles, well past the limits of the curvature of the earth — but he didn't know what.

He was a strange man, Marconi. For one thing, he was the only one of the electrical pioneers who was born rich. No one ever heard him laugh. His smile struck some as unpleasant, and others never saw him smile at all. He had a temper, and he had his quirks: he disliked riding in an elevator with someone he didn't know, although the elevators were the only things that impressed him about America when he finally got there. Alone, too, of the electrical pioneers, he was mostly truthful in his utterances, although he intensely disliked being questioned, especially by the American press. He was not a likeable man, although he commanded the intense loyalty of many who were likeable themselves. He could silence a room simply by standing in it, especially if it was occupied by a woman who was behaving in a way he didn't like. But most women found him irresistible. Like Tesla and Edison, he had an enormous capacity for work. He was uninterested in the transmission of voice or music, and he distrusted any innovation that wasn't under his control. And now, although he would always condescend to the place, he had to go to America to save his company. When there, he would have to do the impossible. He would have to send a radio signal across the Atlantic.

There were gales of unprecedented severity all during the winter of 1901–1902. Marconi's recently constructed towers in Poldhu, Cornwall, and at Wellfleet on Cape Cod were destroyed. R. N. Vyvyan, Marconi's chief engineer, had told him they were badly rigged, but Marconi wouldn't listen. Although new antennae were swiftly jury-rigged at Poldhu, dirty weather slowed the work in Wellfleet. But with money running out, Marconi could waste no time.

He would have to make do with Newfoundland, on the narrowest part of the Atlantic, where Field and Thomson had landed their cable forty years before. Broadcasting to Newfoundland meant that because of the curvature of the earth he would have to overcome a mountain of oceanic water a hundred and fifty miles high. There were no broadcast towers in Newfoundland, but he would have to get his antenna aloft somehow. Equipped with a weather balloon and four huge kites made from a design by Sir Robert Baden-Powell, the hero of the Boer War and the founder of the Boy Scouts, Marconi and his assistants arrived at Glace Bay on December 9. They set up shop on the aptly named Signal Hill. In Cornwall, Edison's former chief British associate, John Ambrose Fleming, manned the transmitter. The balloon was launched, carrying the aerial aloft, and was immediately carried away by the wind.

In Cornwall, by prearrangement, Fleming broadcast every day at ten-minute intervals between 3 and 7 P.M., Greenwich Mean Time. The repetition of the letter S — three dots — was chosen because a dot consumes less electricity than a dash. Marconi sent up a kite, and it followed the balloon into oblivion. He sent up a second kite. It reached an elevation of four hundred feet, and there it remained. But the kite kept bouncing up and down, which caused another problem. Marconi had a primitive tuning apparatus, appropriated from Oliver Lodge, but a tuned circuit was worthless with an unstable antenna. Marconi shifted back to older, untuned equipment. As his receiver, he used a telephone earpiece, thought to be the most sensitive instrument on the planet.

Marconi was, his associates recorded, utterly calm. At regular intervals for two days, he sent the same message to Fleming: "Nothing."

"I am sending," Fleming responded.

On December 12, he sat as usual, composed and with the earphone in place. He picked up his notebook and wrote a short sentence, recording the time. Then he passed the receiver to his assistant.

"Do you hear something, Mr. Kemp?" he asked.

Faintly, three dots, or so they said: the letter S in the code that Vail invented and Morse named for himself.

It was several days before the press was called to Signal Hill to receive the news. The news, in turn, was received with skepticism. No one but Marconi and his men had heard the signal — if signal there were. Everyone knew of Marconi's financial difficulties, and everyone knew that Marconi's associates were a famously loyal lot. Edison, at fifty-seven the world's most senior spokesman on things electrical, was consulted.

* * *

The great man's reaction was interesting.

Marconi, like Tesla, was precisely the kind of person Edison instinctively despised: cultivated, European, and not Edison. But Marconi had a singular virtue that was not apparent to most observers, although it was uppermost in the Edison mind: Marconi wasn't Tesla and — this is a guess, but a good one — he could be used as a weapon against a man Edison had come to hate.

Edison was one of the few people in the world who knew to a certainty how brilliant Tesla was, he knew that Tesla was up to something at Wardenclyffe, and he strongly suspected that, unless he could head Tesla off, the Serbian scientist would steal another march on everyone. But a single word from Edison, properly chosen, could cut the ground from beneath Tesla's feet; old J. P. Morgan, for all his many virtues, knew absolutely nothing about electricity. On the subject of Tesla, Edison was slightly ambivalent — when Tesla's lab burned down in the 1890s, he commiserated with him and lent him space at West Orange — but Edison's ambivalence was not merely slight. Usually it didn't exist at all. Most of the time, he seemed to hate Tesla, and the two of them did not speak. Tesla had robbed Edison of his company and his electricity. Getting Tesla crosswise with Morgan would be Edison's revenge.

Edison knew perfectly well that Marconi could be lying about his Atlantic transmission. He also knew Marconi had an excellent motive for lying about it, because he had to show his investors something wonderful. Like everyone else who knew anything about the subject, Edison firmly believed that Hertzian waves traveled in a straight line. But he also believed that Tesla, given enough time and money, was capable of producing a wireless telegraph unlike any the world had seen or dreamed of.

"If Marconi said he did it," Edison stated forcefully on the basis of no evidence whatever, "he did it."

Then he sold Marconi his useless wireless patent for a form of radio that had no commercial prospects.

* * *

But Marconi wasn't lying — or, perhaps, he wasn't exactly lying. Wireless telegraphy was perfectly capable of crossing the Atlantic, as he proved the following year when, reporters in tow, he duplicated the results with much more sensitive equipment aboard the ocean liner *Philadelphia*. But there's the rub.

On Signal Hill, everyone remarked on how rudimentary Marconi's equipment was even by the standards of the day. And in all the other tests — Judge Vail's at Speedwell, Popov's at St. Petersburg — the recipient had no idea what the message was. Marconi knew exactly. He also knew exactly when it was supposed to arrive. The northern latitudes are extremely noisy during that time of year, and a telephone earpiece, despite what Marconi believed, was precisely the wrong kind of equipment. Almost no modern electrical engineer believes that Marconi heard England in 1901.

In the end, of course, it didn't matter, because radio works the way Marconi believed it did. He never developed an interest in broadcasting voice and music; he was in the business of equipping oceangoing ships, and they communicated in Morse code. In need of a new signal detector to replace the crude Branly coherer and improve his tuning, he put Professor Fleming on the job. Fleming, who'd worked for Edison, remembered the Edison Effect; he'd studied it extensively and had come to no better conclusion than Edison had. He also remembered that there was something mighty peculiar about the way those light bulbs began to work if you put a second filament into them. They seemed to detect electricity. If they could detect electricity, they could be made to detect radio. Thinking about them and tinkering, Fleming invented the modern vacuum tube, the basis for half a century's advances in electronics until the transistor was invented in 1947. John Ambrose Fleming became Sir John Ambrose Fleming. With typical conservatism, Marconi, who became a marquess, did not install the new detectors until 1912, turning his system into a modern, if narrow (because it broadcast only telegraphy, not voice) radio system. As certain shrewd observers, including Tesla, had noted, Marconi's original equipment was a dead end that went nowhere.

The British and the Americans took his broadcasting stations away from him during the First World War. Marconi was the citizen of an allied power, Italy, but the war had made radio so important that it seemed unwise to allow a major chain of stations to remain in the hands of any foreign national. The Marconi Company became a manufacturer of radio equipment. In Britain the stations became the basis for the BBC. In America they became the Radio Corporation of America, RCA, soon to be headed by Marconi's devoted disciple David Sarnoff. Marconi was married, but neither happily nor well; the union did not last. When Mussolini came to power, he became a member of fascism's Grand Council. He always claimed that he was Italy's first fascist, because he had given the name *fascisti* to the short waves he studied during his declining years. It was his little joke. He died at sixty-three in 1937.

To all intents and purposes, Tesla died in 1901 and was buried in

1943. There was one last invention in him: a bladeless turbine that was forty years ahead of its time and generated electricity from water that was little more than sludge, but when forty years had passed and it was finally possible to forge the metals that would make it work, it was obsolete; better turbines had been invented. He often spoke of other inventions. As he grew older, he would call in the press on his birthday and describe some of them, especially the death ray that he was soon to give the world, which would make war impossible. Everyone was making death rays then; Trotsky warned the West not to meddle with his country, because it had a particularly good one. Tesla proposed to run a motor with cosmic rays and never did it. He claimed to have developed a radical new method of purging metals of harmful gasses, but when somebody loaned a lab to try it, he demonstrated that he knew nothing useful about metal. He never gave up his dream of a world broadcasting center using electrical currents as the carrier wave, but he never built it and nobody else did either. He went to his grave believing Marconi was on the wrong track, because radio was not the wave of the future. Neither Colonel Astor nor Morgan ever saw their money again, and Tesla never wore the same pair of gloves or collar on two consecutive days. He was still the sharpest man on Fifth Avenue. But after 1901, except for the bladeless turbine, he never did anything of importance again. When his belongings were inventoried in 1943, the box housing his death ray was finally opened. It contained a Wheatstone bridge, a commonplace laboratory device for measuring resistance across a circuit that was invented by a member of the Christie auction family but popularized by Charles Wheatstone. It was not a death ray, and one was never found.

CHAPTER TEN

VOX HUMANA

The little things of today may develop into the
great things of tomorrow.

— *Sir John Ambrose Fleming*

In the courthouse square of the rural town of Murray, Kentucky,
there stands a statue of Nathan B. Stubblefield, melon farmer,
shack dweller, and mystic. The inscription reads: Father of
Radio.

Actually, he invented the cell phone.

He was born in Murray, and he seems to have left it only once.
There was no mysterious side to Nathan Stubblefield. Everybody in
town knew what he was doing.

In 1892 he handed his friend Rainey Wells a box and told him
to walk as far away as he liked. When his friend stopped, the box
said, "Hello, Rainey." Like Preece and Edison, Stubblefield appears
to have been using induction fields. Rainey Wells became the first
recipient of a call on a portable phone. Its range was about a mile.

In 1902 Stubblefield invited a reporter from the *St. Louis Post-
Dispatch* to bear witness to his latest invention. The reporter was
given a telephone attached to a pair of four-foot metal rods, although just what this device looked like cannot be learned, because
the archive on Stubblefield is very thin. Like Rainey, the reporter

was instructed to walk as far away as he liked and then drive the rods into the ground. The reporter did as he was told. As clear as a bell, he then heard the voice of Stubblefield's son. Like Tesla, Stubblefield was making use of the conductive powers of the earth. The range was about eight miles.

The demonstration won him an invitation to Philadelphia, where he exhibited his phone at the Franklin Institute, and to Washington, where he embarked on the steamship *Bartholdi* with one of his phones, handed others to spectators, and told them to go to a place on the bank of their own choosing. Once again, his voice came across loud and clear. Like Dr. O'Shaughnessy, the East Indian telegrapher, he was making use of the conductive powers of water.

Just what happened then remains mildly unclear. He met some Eastern sharpies and an enterprise, the Wireless Telephone Company of America, was formed to market Stubblefield's device. Stubblefield wrote the preface for the prospectus and filed an incomprehensible patent. Then, embittered by something, he went home. As usual in such cases, he claimed his partners had tried to cheat him, but it's hard to know what really happened. Stubblefield was a peculiar man with some very odd thoughts. He believed he was a witch. He did not seem to crave human contact, and he kept almost entirely to himself. His neighbors saw him from time to time, moving at a distance. Sometimes they detected sounds and lights, but it was impossible to know what he was doing. Echoing Tesla, he had once claimed he could illuminate the mountainsides. The rest of the world had forgotten him.

He starved to death in 1928.

So who invented radio, really? Well, nobody really invents anything. Instead, people assemble them, and there are usually five or six diligent strivers working on the same device at any given time; Alexander Graham Bell and Elisha Gray arrived at the Patent Office within hours of each other. But if we regard the inventive process in terms of the national myth of the lone genius created by Morse and Edison, Tesla invented radio in its basic form. He demonstrated a working transmitter and receiver in 1893, well in advance of anybody else. He understood tuning well enough that in 1898 he used it to build the world's first robot, a remote-controlled boat. Then he went haring off in the wrong direction and left the field to Marconi. Marconi, too, went in the wrong direction. He was interested, to the exclusion of almost everything else, in ship-to-ship and ship-to-shore transmission of Morse code, and his equipment was all wrong.

There is no longer a single commercial Marconi radio set in use anywhere in the world. While its technology was slightly improved by means of Fleming's vacuum tubes, it was extremely clumsy technology that led nowhere. It could not broadcast voice. Its broadcasting apparatus was of tremendous size. It used extremely long-wave transmissions, while the upper end of the broadcast spectrum, where the waves grew progressively shorter, was a much better transmitter although it got progressively harder to use.

Marconi was using a pickup team of other people's components, he was indifferent to theory, and an indifference to theory meant that he could not discern the future of wireless broadcasting. As power-distribution experts, Tesla and Reginald Aubrey Fessenden, Edison's former chemist, lived and breathed theory, because theory told them where the next step in their investigations might lead them. Theory told them that Marconi's technology not only was dead-ended but, absent a monopoly, was commercially unviable, because when radio went to the next level, Marconi would either have to abandon his technology or founder. Indeed, Marconi's company was perennially strapped for cash even when it was a monopoly. In the end, Tesla's was led by his own theories in precisely the wrong direction. Reginald Aubrey Fessenden's grasp of theory led him in the right one. If Tesla's early, correct experiments led him to invent the basic form of radio, Fessenden's theories, which he properly followed up, led him to invent radio in its modern form. Forgotten is the name.

Unlike Edison, he had no sense of publicity and its uses. For his reputation and standing with posterity, this was a catastrophe. His only biographer was his wife, and her book has never been widely read. It was the Nathan Stubblefield problem all over again. Almost no one has ever heard of Reginald Aubrey Fessenden, although everyone listens to his radio. Like Elihu Thomson, he was one of the greatest scientists ever to spend his life and afterlife in almost total obscurity, and like Stubblefield's, his archive is very thin.

Like Tesla, Fessenden realized that Marconi was using a dead-end technology and the end would come soon. Tesla and Fessenden realized that the future resided in voice transmission. And unlike Tesla, Fessenden got it exactly right.

He was Canadian born, the son of an Episcopal priest, and it was clear from an early age that he was flat-out brilliant. Later it was said of him that he was arrogant, irascible, insufferable, and that he richly deserved the obscurity into which his name has fallen despite the fact that his subsequent inventions, such as strategic bombing, the bombardment of an enemy country by fleets of warplanes, won what Winston Churchill called the wizard war of 1939–45. He was none of these things. Instead, he was a man who thought with incredible speed; Fessenden thought faster than most people who ever lived. He graduated from Trinity College School in Port Hope, Ontario, at the age of fourteen. Still a teenager, he was given a teaching mastership at the Bishop's College School in Quebec. He then moved to Bermuda to become the headmaster of his own school, of which he was also the only teacher. It was not a good career move, although he found a wife there. Bored to tears, he entertained himself by reading back issues of the *Scientific American*, and he soon decided to go to New York and take up the life of a writer. At this point his training was in literature, the classics, languages including the inevitable Greek, and math. In New York, his credentials did not impress. He got the occasional freelance assignment but nothing else, and he decided to go downtown and hook up with Edison. He sent in his card. Edison, who was not having a good day, sent out a note: "What do you know about electricity?" Fessenden, as it happened, knew a certain amount from reading the *Scientific American*, but he decided not to risk his luck with the great man. He wrote back: "Nothing." A second note from the invisible Edison materialized: "I already have enough people who know nothing about electricity."

But, by dint of sheer hanging out in that year of 1886, when he was all of twenty years old, Fessenden got himself hired by Kruesi, Edison's machinist, as a tester. Edison, having finally separated his backers from another sum of money, was laying his midtown wires then, and a tester's job was to test them for faults. When a fault was found, the tester's job was to keep a wary eye out for the police, who would expect a bribe in the form of a fee, get up the pavement, repair the fault, and relay the street in a flawless, none-the-wiser manner. Fessenden was extremely good at all these tasks. He was soon chief tester. Soon, too, he came to the attention of Morgan. Morgan had two serious gripes with Edison's electricity. For one thing, he never believed his bill and continually disputed it. For another, his house kept catching fire. Fessenden soon discovered why. Morgan's wires were insulated with rubber and, following common practice, were embedded in the plaster of the mansion's walls. Fessenden recommended that they be removed from their hiding place and covered with armored cable to protect the rubber, and this became an industrywide practice. Fessenden, who understood insulation as Edison did not, went on to become the inventor of insulated electrical tape. Morgan was impressed with the lad and conveyed his impression to Edison. Soon Fessenden was working with Edison as a glorified office boy.

"Fezzy," Edison asked one day, "do you know anything about chemistry?"

"No," Fessenden said. This time, it was the right answer.

"Then I want you to be a chemist. I've had a lot of chemists. I had one whose name was all through Watt's Dictionary [the bible of the chemical profession]. But none of them got results. I want you to take it up."[1]

By the time he turned twenty-four, Fessenden was not only one of Edison's chemists, he was Edison's chief chemist. On his lunch breaks, he studied math with Arthur Kennelly, another completely self-taught man. He asked Edison's permission to look into the new Hertzian waves that had been discovered in Germany. "Fezzy," said the great man, "that'll happen about as soon as a man jumps over the moon." Still, he gave Fessenden an opportunity to look into it, but not until Edison got back from his trip to Paris, where he was greeted with great acclaim. And by then it was too late.

Fessenden would beat Edison's prediction about the man and the Hertzian waves by sixty-three years, but he didn't do it on Edison's time. Edison's pocketbook took a huge, unexpected hit when Villard consolidated the enterprise (and sold it to the Germans) as the Edison General Electric Company. The chemistry department was abolished in the subsequent austerity campaign, and Fessenden was out of a job. Not for long, though. Soon he was in Newark, working for George Westinghouse. While there, he produced the two inventions that made it possible for Westinghouse to use the Sawyer-Man light bulb in Chicago by making it nearly airtight, evading the Edison patent and bringing the wonders of alternating current to the attention of the great American public. Fessenden next decided to become an academic, first running the new electrical engineering department at Purdue and then, at Westinghouse's urging, taking an engineering professorship at the University of Western Pennsylvania in Pittsburgh. But the academic life could not hold him, and he was soon happily, and then very unhappily, at work for the Weather Bureau, which was then establishing a chain of wireless stations up and down the East Coast as a replacement for Joseph Henry's telegraphy, to try to anticipate the weather it reported.

At this point, as the century turned in 1900, two things began to make themselves apparent in Fessenden's life. First, he was finding it increasingly difficult to get along with people who were important to his career, although it is difficult to know why. If Tesla is badly served by his biographers, Fessenden is hardly served by them at all. He never completed his autobiography, in part because it became too dull to print. His chief biographer was his wife, an understandably partisan woman who saw matters through his eyes, and in

Fessenden's eyes he was suddenly beset with enemies on every hand. Perhaps it was so, but one suspects an unworldly streak in the man. He never learned anything useful about business, and he was extremely bad at it. Unlike Edison, Tesla, and Morse, he had no gift for self-promotion. He did not have Edison's way with words, and he lacked Edison's common touch; Tesla's accomplished showmanship was a stranger to him as well. And he had the unfortunate habit, commonplace in men of extraordinary intelligence, of assuming that everyone else was as smart as he was; much of the time, to most people, he was completely unintelligible. Perhaps it was some combination of these qualities that led him to perceive villains all around him. Perhaps he was right. Or perhaps it was an early onset of paranoia.

And, at this point in his life, there was a second thing about him. He had realized that a dynamo, if run in a certain way that was almost impossible, could become a radio transmitter. That is, it would generate radio waves, and a radio wave, as Marconi was crudely demonstrating, was a carrier wave; in Marconi's long-wave system, the wave carried information in the form of a Morse dot or dash. But Marconi operated his equipment using a whiplash effect. He turned it on, effectively introduced a short circuit into his system that made the spark leap across the Hertzian gap, recorded the short as a dot or a dash, turned the equipment off, turned it on again, and repeated the whole process. Indeed, Marconi and his scientific advisor, John Ambrose Fleming, believed that this was the only way to operate a radio, and Fleming wrote an immensely influential text on the subject. As long as this delusion persisted, voice transmissions were impossible; the spark was too noisy, and with a top capability of six sparks a second, it was too slow. Six sparks a second wasn't enough to spell many words per minute. It was Fessenden's insight — and Tesla's — that alternating current could become the carrier wave. Alternating current was continuous, not whiplash, and the wave itself could be shaped into an analog of the human voice, which is how the trick of modern radio is done, because sound is also a wave. If you changed an alternating-current radio wave into an exact duplicate of a voice wave, a voice would come out of a receiver tuned to it. Tesla's mistake was to choose ground transmission while toying with a version of the old Mahlon Loomis system developed by the Virginia dentist at the end of the Civil War, using the conductivity of the upper atmosphere discovered by Franklin, a conductivity that does not exist in the absence of cloud cover. Marconi, Fessenden realized, had gotten one thing right: once you generated your radio waves, you had to send them through the air, whether it had clouds in it or not.

As a young man at Edison's knee, Fessenden had learned certain things about the art of experimentation. You did not stop when you got a good result. You kept right on until all possibilities were exhausted; your seeming good result might actually be a bad result, an indifferent result, or no result at all. You moved by increments, doing what was possible *right now*. You did not design components; you designed entire systems of which the components were the parts. And as long as you knew where you were going, you never stopped.

Because it was impossible to build a dynamo to his highly demanding specifications, Fessenden did the next best thing. He obtained a common Edison phonograph cylinder and had it milled with ten thousand microscopic horizontal grooves. Introduced into his system to control the flow of electricity, it would generate ten thousand sparks a second and give him a facsimile of a bad carrier wave, but one that would convey the human voice. Using it, he climbed a fifty-foot radio tower on Cobb Island in the Potomac on December 23, 1900, and spoke into a microphone. "One, two, three, four," he said. "Is it snowing where you are, Mr. Thiessen? If it is, telegraph back and let me know." It was an absolutely terrible transmission, but it worked. The Cobb Island experiment was the first broadcast of radio as we know it, although the public did not learn of it. Fessenden had passed up his first chance to join Franklin, Morse, and Edison in the ranks of the immortals.

But things were not going well in Fessenden's life. The Weather Bureau was no longer delighted with Fessenden, and Fessenden was no longer delighted with the Weather Bureau. Some sort of colossal misunderstanding seems to have erupted. Fessenden's contract specifically stated that he owned any inventions developed on the bureau's nickel, but he had somehow become convinced that the bureau's chief not only wanted to own, in his personal capacity, all of Fessenden's patents, but he also wanted half of the money Fessenden earned from them. That can hardly have been so. The contract existed, and all Fessenden had to do was produce it. He could destroy the career of the Weather Bureau's chief, and the chief, unless he was an exceptionally stupid man, had to know it. But Fessenden didn't seem to understand the law of contracts very well. After firing off a letter of protest to President Theodore Roosevelt and getting no answer, he quit.

He had no savings to speak of. He and his wife had to borrow on their life insurance, but his lawyer managed to hook him up with a couple of Pittsburgh millionaires, T. H. Given and Hay Walker, and it looked like plain sailing for a new enterprise called the National Electric Signaling Company, Nesco for short. But once again, Fessenden was fooled by a huge misunderstanding. The Pittsburghers, who eventually earned Fessenden's undying enmity by pouring two million dollars into Nesco, thought they were buying a completed technology. Fessenden, who knew that years of experimentation lay ahead but was no good at explaining himself to a layman, thought they understood they were buying a process, not a product. He also didn't understand that even a millionaire's patience could wear thin, especially if the millionaire had no idea where his money was going. This compounded the problem further and pretty much left Fessenden to his own devices, as though he could actually sell a product when he finally made one. He couldn't. He had no idea how.

He tried. As the result of a happy lab accident, he designed a marvelous new receiver, the best in the world, and using his linguistic skills named it the barretter after the French. It was simplicity itself: some induction coils, an electrode submerged in a conducting liquid, and a pair of headphones. It was an interim de-

vice — his radio-generating alternating-current generator was the
ultimate goal — but, to add to his mounting feelings of persecution,
his patent was immediately infringed by the United States Navy,
which built barreters for itself and refused to pay him. No doubt the
Pittsburghers, both accomplished men of affairs, would have known
how to deal with the problem, but the Pittsburghers gave no indication of knowing that a problem even existed. Having neglected
to tell them but still spending their money, Fessenden set up a headquarters in Washington and built a four-hundred-foot antenna in
Brant Rock, Massachusetts. Next, in 1906, he approached Charles
Proteus Steinmetz, the presiding genius at General Electric's Schenectady works. He wanted Steinmetz to build an alternator — a
dynamo — that revolved at an unprecedented rate and produced
an unprecedented radio wave. Steinmetz tried, but even a genius
could not defeat the laws of physics. The metallurgical technology
simply did not exist to build the device Fessenden wanted, although
a device of some sort was eventually delivered. Fessenden, who was
considerably smarter than a genius, rebuilt it in the machine shop
he'd equipped thanks to the generosity of Given and Walker so that
it almost produced the waves he needed. Then he sent the plans
showing what he had done back to Schenectady and asked Steinmetz to try again.

Steinmetz gave the job to a young Swedish engineer named
Ernst Alexanderson, who was so absentminded that he sometimes
shook hands with his own daughter and had been seen in conversation with a telephone pole. It had become a common practice for
recently graduated Swedish engineers to spend a couple of years in
America, the hotbed of technological change. Outside of Germany
and Hungary, innovation had grown so moribund in Europe that
the new power plants in South Africa and India were being built by
Americans and Canadians. Thanks to Fessenden's alternator,
Alexanderson carved himself out a career as a General Electric inventor, remained in America, grew rich and powerful, and became
a citizen, although his brother engineers thought Fessenden's alternator would kill him the moment he tested it. It would fly apart,

they said, and impale him between the eyes. Just to be on the safe side, Alexanderson dug a hole, sandbagged it, and tested the alternator there. It worked. For years afterward, until the vacuum tube ascended to the throne, the Alexanderson alternator was the basis for radio. And for years afterward, Fessenden wrote General Electric and pointed out that because he, not Alexanderson, had designed the revolutionary device, it should be called the Fessenden alternator. GE always wrote back and said they would look into it. But nothing ever changed.

With the alternator in hand, and again apparently without consulting his backers, Fessenden decided to get into transatlantic broadcasting, and in 1903, a year after Marconi's probable first such broadcast, he built a second station in Scotland. It was he, not Marconi, who made the first two-way communication between America and Europe. His apparatus was so sensitive that, under certain extremely peculiar atmospheric conditions, Scotland could hear Fessenden's assistant, Adam Stein, talking to a small satellite station ten miles away from Brant Rock. Here, indeed, was a newsworthy item; it would be the making of the Fessenden name and fortune, and he invited his old friend Elihu Thomson, Edison's inventive rival and Steinmetz's predecessor as the presiding genius at General Electric, over for a demonstration. Then the Scottish tower blew down in a storm. Until it was rebuilt, there would be no broadcast, and it could not be rebuilt in time for an important demonstration he had scheduled. Fessenden would have to think of something else.

On Christmas Eve 1906, with Elihu Thomson and representatives of the telephone company present and observing, Fessenden plugged in his alternator. At nine o'clock in the evening, he took his telegraph key and tapped out "CQ CQ CQ" — the signal indicating that he was about to broadcast — in the trained telegrapher's rapid sequence. It was the general call to all stations within range. Every wireless operator on every ship within a hundred miles — most of them belonging to the United Fruit Company, which was just about the sole paying customer for Nesco's radios, but some of

them belonging to the navy, using pirated Fessenden barreters —
leaned forward to transcribe.

Fessenden picked up his microphone and explained that a test
of voice transmission was now under way. He then played a rendition of Handel's Largo on an Edison phonograph and turned to his
assistant.

"Mr. Stein?" he said.

The first case of mike fright in human history occurred. Mr.
Stein was unable to utter a word. Neither were Fessenden's wife and
secretary, who were scheduled to sing. Fessenden swiftly saved the
day by snatching up his violin and launching into "O Holy Night."
Then he wished everyone a Merry Christmas, announced that a
second test would be held on New Year's Eve, and signed off.

Unbelievably, he forgot to tell the press. Once again, the public
did not know that voice broadcasting had been achieved, or by
whom.

Still, Elihu Thomson and the representatives of the telephone
company were deeply impressed. Unfortunately, Fessenden was a
customer of Thomson's company, not a supplier, and GE saw no
compelling reason to go into the radio business. The people from
AT&T did. They were in the voice transmission business, radio was
clearly the Next Big Thing, and Fessenden had the necessary radio.
In fact, it would probably be dangerous to ignore Fessenden's radio,
for reasons that were rather obvious. The Fessenden fortune was
made.

It wasn't. It probably would have been if Charles Coffin, the former shoe manufacturer and former head of General Electric, had remained at the AT&T helm, but Coffin departed, and with Coffin's
departure, the thirst for innovation went out of AT&T, no matter
how enthusiastic its witnesses at Fessenden's demonstration were.
Fessenden did not possess the one quality Edison had in abundance:
luck. In the vicinity of Reginald Aubrey Fessenden, if something
could go wrong, it would. Not only did Coffin leave his post, but the
trust company panic of 1907 — Morgan's finest hour and last hur-
rah — badly panicked his successor, who was, by a supreme irony,

Theodore Vail, a descendent of Alfred Vail. In 1907, AT&T was not the innovative pioneer that corporate legend makes it out to be. AT&T hardly innovated at all; it preferred to buy its components off the shelf and cheaply. In corporate terms, Fessenden's radio was not expensive, and Vail's engineers were loud in its praises, especially when they compared it to the clunker Marconi had built. But Vail was not an engineer, and there was the panic on Wall Street. Vail passed on one of the most remarkable opportunities ever presented to an American corporation. AT&T, a legal monopoly, could have owned radio and all that came after.

For Fessenden, the end was nigh. Nobody knew he'd invented the radio, and the people who knew didn't want it. He hung on for a while, never ceasing to improve his creation, and then his investors pulled the plug. They were decent enough fellows and they genuinely liked Fessenden, but they were also fond of their money. And they were extremely puzzled by the fact that neither Fessenden nor his wife seemed to understand the contract he had signed, giving Nesco the ownership of the radio patents; Fessenden, in fact, was furious, which might explain why two decent fellows from Pittsburgh, Given and Walker, waited years before they settled out of court for half a million dollars that they didn't miss. They were, after all, wealthy men. Fessenden never forgave them, although his own spectacular blunders, not theirs, prevented him from being known, right down to the present time, as the man who took radio to the next level.

Unlike Tesla and Edison, he didn't go strange, and unlike Morse he didn't quit. He developed an iceberg detector, an early form of radar. He invented an underwater detection device, an early form of sonar. He developed a wireless direction finder called the pelorus. Fessenden's pelorus is a vital part of every aircraft in the world; although many pilots have no idea where it came from, it is the instrument that tells them where they are. When the First World War rolled around, he offered his devices to the British and the Americans. The British were using his radar against the zeppelin raiders, but neither they nor the Americans bought his underwater detection devices, although the U-boat came close to driving Britain to starvation and out of the war. Fessenden evolved a plan to build ten thousand bombing planes a day in Canadian factories, where the American principles of mass production were understood. Flying in tight, deadly formations, these planes would then carry the war to the enemy, bombing out the sinews of his war machine. The British passed on that, too. Another thirty years would come and go before the skies of the Reich were blackened by Fessenden's formations, and Fessenden would not live to see them. "Don't worry, Fezzy," Edison wrote, for he remained on good terms with his former chemist. "They won't listen to me, either." Edison had finally gotten the Allies to stop making sitting ducks of their ships and form them into convoys, but that was as far as he got. Edison's and Fessenden's inventions won the Second World War, not the First.

Fessenden spent his last years compiling a book on his theory that the Biblical deluge occurred in the vicinity of the Black Sea as a result of glacial melt, and the Black Sea also accounted for the Greek legends of Atlantis. Seventy years later, the Fessenden Black Sea flood thesis (although no one calls it that) has finally found favor. He was the only man Edison ever regarded as his equal.

He developed heart disease and retired to Bermuda, the home of his wife, where he had once been a one-man pedagogical band at the ripe age of seventeen. He died in 1932. On his tombstone, in Egyptian hieroglyphics, is this epitaph: I Was Yesterday, but I Saw Tomorrow.

CONCLUSION

In the second century B.C., the Chinese discovered that blowing air through molten iron resulted in steel, but it took them six hundred more years to figure out how to build a blast furnace. When the Englishman Henry Bessemer rediscovered the phenomenon in the 1850s, he figured out how to build a blast furnace within months. One of the abiding features of the golden age of invention that began in 1769 — the *annus mirabilis* that marks the beginning of the Industrial Revolution, when Richard Arkwright built the first automated spinning mill in Derbyshire and James Watt, a Scots scientific instrument maker, fitted a condenser to an old pump and created the modern steam engine — was its remarkable speed. Another was its duration. In most of human history, inventive spurts lasted for about seventy-five years, followed by a lull, and then invention started up again. China was particularly fruitful in this regard, creating a host of new products and devices — the horse collar, the compass, gunpowder — which the medieval Europeans cheerfully stole, advancing their own somewhat more backward civilization. China was large, unified, and stable; the on-again-off-again inventive cycle flourished there for millennia. But the Western inventive cycle that began in 1769 did not peter out after seventy-five years. It went on and on. It slowed down for a while, starting in the early years of the twentieth century. Then it sped up again in the 1970s, and it's still going on to this day. In the surmised and recorded thousands of years of human history, this has never happened before.

The age of the lone inventor also never really happened, exceptions to the contrary — Edison and the phonograph, Bessemer and
the blast furnace — notwithstanding. Science tends to erupt at certain times, as was the case with the telegraphic explosion of the
1830s and the sudden proliferation of electrical generators thirty
years later, and it is a collaborative, not a solitary affair. Morse had
his assistants and so did Edison; men who worked alone, like Bessemer and Fessenden, were the exceptions. But even the solitaries
were not alone. They were functioning in a certain scientific environment, and science is created by many people. Still, it can be said
that with Fessenden's Christmas broadcast of 1906, the first phase of
the golden age came to an end and was replaced by something else.
Invention went indoors, became corporate, and slowed down, and
it stayed that way for seventy years.

The case of Philo Farnsworth tells the story. He was another
solitary, the lone inventor in the era of the corporate lab who is the
exception (aside from a few relatively unimportant tinkerers) to our
tale. In or around 1920, when he was fourteen, he conceived of
electronic television by contemplating (or so he said) the orderly
rows of his family farm in Idaho. They reminded him, he said, of the
broadcast spectrum, and they may also have suggested the final form
of a television picture, which is made up of lines. As a very young
man just graduating from high school, he attracted some modest
backers, moved to San Francisco, and rented a lab. He never went
to college, but by 1927 he had a working television system. So far,
Farnsworth was following a time-honored American pattern, but he
soon found that he was not operating in the time-honored American inventive environment that had prevailed for the hundred and
seventy years that had passed since Franklin flew his kite. Farnsworth
had enemies. One enemy was a person, David Sarnoff, who was
neither a scientist nor an inventor. The other enemy was the electronics company Sarnoff ran, RCA, and this was something entirely new.

Marconi's American wireless stations had been seized by the
government during the First World War, and at the end of the

conflict it was not deemed prudent to return them, because Marconi was a foreign national, his stations were an American treasure, and some very powerful companies — General Electric, Westinghouse, and United Fruit, which depended on radio to communicate with its ships and Central American banana plantations — wanted them. And got them, forming a new cobbled-together corporate entity called the Radio Corporation of America. Through its owners, RCA controlled virtually all the radio patents in the country. It also had another asset in the form of a brilliant young executive named David Sarnoff, who in 1915 conceived of a marvelous new consumer product that he called the Home Music Box. It was, in fact, the home radio receiver. His superiors were not keen on the idea, and to Westinghouse went the honor of introducing the radio to the American home. But Sarnoff was soon head of RCA, in a position to implement his idea, and presently found himself in command of a changed company. The government had forced out its corporate owners, who had called the shots and were not reluctant to exercise a veto. RCA was now owned by private stockholders, who had very little idea of how the company was run. This gave Sarnoff tremendous power, and the fact that RCA owned virtually all of the American radio patents gave the company tremendous power in the industry. If a competitor tried to roll out a superior product, the RCA lawyers could claim infringement, tie up the competitor in court for years, and perhaps prevail. If a competitor did something else Sarnoff didn't like, RCA could and did threaten to withdraw the licenses to use the RCA patents, putting the competitor out of business; nobody could build radios without RCA's patents. Sarnoff knew just how to exploit such an edge, but he was a Russian immigrant with an immigrant's exalted and even worshipful notion of art and commerce, which made him attempt a couple of things that would seem extremely peculiar to subsequent generations of corporate executives. In partnership with Commerce Secretary Herbert Hoover, he tried to keep advertising off the airwaves. AT&T, which briefly owned radio station W(ind) E(arth)

A(ir) F(ire), rented out time to corporate sponsors, and the camel's nose was in the tent.

For years Sarnoff tried to coax the American public into listening to classical music, maintaining a symphony orchestra at the RCA-owned (and first) radio network, NBC, and hiring the great Arturo Toscanini to conduct it. The American people stayed away from the NBC Symphony in their millions. Nor was Sarnoff particularly fond of vaudevillians like Jack Benny, George Burns, and Gracie Allen, and he thought they were easily replaced. He found out otherwise when William Paley, the head of the up-and-coming CBS network, stole the vaudevillians away and the listening audience followed. But if there was an idealistic and naïve side to David Sarnoff, there was also a ruthless side, and Philo Farnsworth soon felt the kiss of the Sarnoff lash.

Sarnoff, unfortunately for Farnsworth, had taken an interest in television. The concept had been brought to him by another Russian immigrant, Vladimir Zworykin, who had once studied at Popov's Imperial Torpedo School in Kronstadt, where much of the early work on television had been done. But Zworykin had been unable to interest his previous employers at Westinghouse in the idea, although he thought (and told Sarnoff) that a practical television system could be created with a few tens of thousands of dollars. Sarnoff, who spent millions, often amused audiences with the story of Zworykin's comical notions of television's true cost. But Zworykin himself represented something that was new: the corporate scientist-inventor. He worked for RCA, which meant that he worked for Sarnoff. He drew a salary and kept business hours. His job was to invent, not for himself and not for humankind, but for Sarnoff and RCA, and he was given a salaried staff of scientists to help him. This was the new wave in inventing. With Zworykin and others like him, the golden age of invention passed into history.

Even with the huge resources of RCA behind him, Zworykin found that he could invent a good television receiver but not a good television camera. Farnsworth, laboring in San Francisco, had built such a camera, and word had reached the papers. Sarnoff sent Zworykin out to have a look. He hung around the Farnsworth lab for a couple of days, sharing the comradeship of science, and then returned to New York and reported to his employer. "I wish I'd invented it," he said of Farnsworth's technology. That was enough for Sarnoff. He offered to buy Farnsworth's television for a few thousand dollars. Farnsworth preferred to license his technology. RCA never licensed anything. Sarnoff buckled down to get Farnsworth's television away from him.

Farnsworth was a lone inventor, but he was not averse to placing himself under corporate protection as long as he continued to own his patents, and he moved his operations to Philco, a now-defunct Philadelphia electronics company. Sarnoff followed. His agents seduced Philco secretaries in hopes of learning something or laying hands on a useful document. Other agents attended Farnsworth's lectures and demonstrations, hoping he would drop a clue to his secrets. RCA, which owned many of the patents vital to Philco's work, applied pressure, trying to force Farnsworth out of there and into the open, where he would be more vulnerable. Philco capitulated, Farnsworth was sacked, but in the end an adamant Farnsworth won. With Zworykin making no progress, RCA was compelled to buy the rights to his camera. The RCA lawyer wept as he signed the licensing agreement.

Unfortunately, Farnsworth was not destined to be rich. Sarnoff rolled out his new Zworykin-Farnsworth television in the presence of President Roosevelt at the New York World's Fair of 1939. "Now we add sight to sound," he announced. Not so fast. Because the television signal was so large, the only way to create a figure was to paint it in lines on the picture tube — paint it twice, in fact, sending first one half of the picture and then the other half. In simplified form, it happened like this: The interior of the picture tube was imbued with chemicals called picture elements, or pixels, which reacted to only those portions of the television signal that contained a particular pixel's lines and textures. In the rear of the picture tube was an electronic gun that fired forward the television waves, but it did not behave like a shotgun, scattering its waves all over the pixels. Instead, it behaved like a rifle wielded by a skilled marksman. Starting at the top of the picture tube, the electronic gun would fire off one line of the picture, then another line, and then another line. Then it did it again. Mixing their metaphors, television experts call this "painting," and because it was done very rapidly, the viewer saw the picture as whole because of a trick of the human brain called persistence of vision. In other words, the brain could not keep up with the television set, although the photographic camera can, which is why a photograph of a television picture reveals both the lines and the picture's lack of completion. But the number of lines painted on the tube dictates the quality of the perceived picture, and the more lines are crowded onto the tube, the clearer the picture will be, which is why European television, with 625 lines, is sharper than American television, with 550 lines. This will change when digital television, using techniques developed for computers, replaces the current system, but this same system prevails for now. Sarnoff's prototype television had only 440 lines. There was a slight delay while the Federal Communications Commission settled on a mandatory format. There was a longer delay for World War II, during which no television sets were sold. Farnsworth, who had bought a radio factory in the Middle West, could neither sell televisions nor collect on his RCA license, because RCA wasn't making any televisions either; a patent licensee does not pay rent on an idea but only forks over money when it makes something. When television sales exploded after the war, his patents had only a short while to run. His house in Maine, in which he had invested much of his money, burned down. It was not insured. He became an alcoholic. In the last years of his life, he was trying to achieve cold fusion.

He was, it was said, the last of a breed. For most of the twentieth century, most inventors worked in teams at corporations, the corporations owned their inventions, and commentators, falling into the usual human tendency to regard current conditions as permanent, believed this was the way the inventive process was always going to work. Moreover, the inventive process had slowed down, relatively speaking; compared to the rest of human history, the inventive process was running at a blinding clip, but compared to the golden age, there was a perceptible falloff. In part this was because, remarkable and complex as the inventions of the golden age had been, most of the easy inventive science had been done; it was now much harder to invent things. In part, however, the slowdown was in the nature of the corporate beast. Teamwork slowed things down; there were too many opinions to consult. Paperwork slowed things down. Bureaucracy slowed things down. And companies owned inventions they didn't know they owned. There is now a cottage industry in plowing through corporate patents to find the things that were invented but fell through the cracks. Then, in the late 1970s, things suddenly started going fast again. The age of the lone inventor had returned.

Here, the tale of the Two Steves, Wozniak and Jobs, tells the story. It begins at Xerox PARC, the copier company's Palo Alto Research Center in California. Xerox PARC, like many corporate labs, was designed to be an invention factory, and it was a very good one. It had, in fact, invented the desktop computer, which it called the Alto. It had invented the mouse. It had developed object-linked computing — pointing with the mouse at a picture on the screen that represented, say, the attached printer and clicking on it, whereupon the printer would spring to life and reproduce the words or images that occupied the rest of the screen — to a high art. A single Alto cost a whopping $12,000, and Xerox, which had never sold a computer in its corporate life, didn't know how to sell it. Steve Jobs thought he could.

Jobs, a college dropout who was designing computer games for Atari, was very interested in the Alto. He would describe its features to his friend Steve Wozniak, who did have a college degree and was working at Hewlett Packard as an engineer. Wozniak then designed a cheap Alto clone and, working in Jobs' parents' garage, they built one and named it the Apple. With the Apple's success, lone inventors began to emerge from garages and basements all over the Bay Area. The computer industry was born in its modern form.

IBM soon got into the act with its own corporate design team, and designed its own desktop computer with off-the-shelf components because its design team had failed to catch the wave of the future and had not designed a proprietary computer protected by IBM patents. Because almost every computer company then copied the IBM design, IBM's nonproprietary (they didn't own it) design came to dominate the industry, although IBM, running in the middle of the computer pack, did not. The age of the personal computer was born. Inventing was up to speed again. Bill Gates cofounded Microsoft to design software for the new industry, and he became the richest man in the world. And as he was becoming the richest man in the world, he began to worry about something.

On the wall of Gates' office in Microsoft's Redmond, Washington, headquarters, there was a picture of Henry Ford. It was not there because Gates admired Henry Ford. It was there because Henry Ford made him worry. At the apogee of Ford's power in the 1920s, his company manufactured half the automobiles in the world. The source of this power was the Model T, and Ford had ordered that the Model T could not be changed. This worried his son Edsel and the forward-thinking executives that Edsel had gathered around him. Over at rival General Motors someone was thinking, as the saying went, outside the box. GM was introducing innovations like automatic windshield wipers, electric starters, and heaters, and it was picking up customers. And so, while Henry Ford was visiting Europe, Edsel and his team designed and built an improved Model T. When Henry returned, they unveiled it. Henry stared at it wordlessly. Then he ripped off one of its fenders. He ripped the top. He took off his shoe and began using it to rip the upholstery. His message was clear. The Model T was not to be changed.

Henry Ford was crazy, of course. It happened suddenly, in 1916. Until then, he had been the model employer who had introduced the revolutionary wage of five dollars a day. His company had been the most benign example of corporate capitalism on earth, with free social services that exceeded the government's, which, in truth, weren't very much. He had been a visionary industrial pioneer who had revolutionized the factory floor with the assembly line. Now all that came to a halt. He spied on his employees and hired goons to break up their unions. He wanted things to change at his own pace and not at a pace that matched the march of automotive science. Like Edison, he began to pay his bills by weighing them. Ford Motors never manufactured half the cars in the world again. This was what worried Gates.

As Gates became the richest man in the world, his company had become commensurately large. The software was designed by individuals working in teams, and it was full of bugs. Microsoft was probably missing something out there that was made possible by new science. It was probably failing to innovate in-house. It was centered on a single individual, Gates himself. It manufactured 80 percent of the most important software in the world, the software that made the dominant non-Apple computers work. Microsoft was becoming like Ford. Or perhaps it was like Xerox, a company that designed a revolutionary device and had no idea how to sell it. But Gates had glimpsed a possible solution. His friend Steve Ballmer, a former Procter and Gamble man, knew how to sell things. Fine, let him sell them. Gates resigned as the Microsoft chief executive officer and put Ballmer in his place. He retained the chairmanship for himself, but he got out of the way. He had seen the future once, when he helped design the essential software that took the world by storm, and perhaps he could see it again. He appointed himself, in effect, Microsoft's chief visionary. But the future was hard to read.

The return of the lone inventor was not an unmixed blessing. A computer-related feeding frenzy on Wall Street created a bubble economy in advanced technology, and bubbles exist to do only one thing: to pop. When this bubble economy popped, it took down good products and ideas with it, along with a lot of bad ones. This was not necessarily a bad thing. The first modern bubble, which occurred in England during the 1720s, was called the South Sea. When it collapsed, it took a portion of the royal fortune with it, and a law was passed forbidding the creation of joint stock companies without parliamentary approval, because the South Sea had been a joint stock company. But the royal family was not wiped out, and the Industrial Revolution came along forty years later. We are now in the Third Industrial Revolution. The first was powered by steam, the second by electricity, and the third is all tangled up with a bunch of computers. So what's next? Well, what about quantum computing? A quantum computer, when it is built, will be able to read every secret code in the world — an example of the kind of power it can deploy, a power unseen by humankind at any time in history. Add stem cell technology. The secret of stem cells, hiding in plain sight, is that they can bestow eternal life. Those are only two examples of a future waiting to happen. There are others, and there is also a constant in this kind of thing. Whatever happens, we will be taken by surprise. We always are. That's how the future works.

ENDNOTES

INTRODUCTION

1. Duane H. D. Roller, *The* De Magnete *of William Gilbert*, 99.

CHAPTER ONE

1. Catherine Drinker Bowen, *The Most Dangerous Man in America*, 60.
2. Leonard W. Labaree, ed., *The Papers of Benjamin Franklin*, vol. 4, 19–20.
3. Ibid., 367.
4. Al Seckel and John Edwards, "Franklin's Unholy Lightning Rod," 4.
5. Bowen, 62–63.
6. Labaree, vol. 3, 127.
7. Ibid., 131–32.
8. Michael Fowler, "Historical Beginnings of Theories of Electricity and Magnetism," 4.
9. Ibid., 10.
10. H. W. Brands, *The First American: The Life and Times of Benjamin Franklin*, 193–94.

CHAPTER THREE

1. Harland Manchester, "The Blacksmith of Brandon," 121.
2. "Bright Lamps, Bold Adventure: A Hole in the Floor," 1.
3. Joseph C. Michalowicz, "Origin of the Electric Motor," 1037.
4. Samuel Rezneck. "The First Electric Motor," 30.
5. Nathan Reingold, ed., *The Papers of Joseph Henry*, 2: 445–46.
6. Rezneck, 30.

CHAPTER FOUR

1. Carleton Mabee, *The American Leonardo*, 149.
2. Ibid., 181.
3. Neal McEwen, "Morse Code or Vail Code; Did Samuel F. B. Morse invented the code as we know it today? Franklin Pope and William Baxter give some answers," 2.
4. *Journal of the Franklin Institute in the State of Pennsylvania and Mechanics Register* 25, 106–8.
5. Mabee, 209.
6. Ibid., 217.
7. Reingold, 3: 216.
8. Mabee, 231.
9. Reingold, 97.
10. Cam Cavanaugh, Barbara Hoskins, and Frances D. Pingeon, *At Speedwell in the Nineteenth Century*, 53.
11. Ibid., 2.

CHAPTER FIVE

1. Samuel Carter, *Cyrus Field*, 97.
2. John Mullaly, "The First Atlantic Telegraph Cable," 13.
3. Bern Dibner, *The Atlantic Cable*, 13.
4. Philip B. McDonald, *A Saga of the Seas*, 70.
5. Carter, 230.

CHAPTER SIX

1. Frank Lewis Dyer and Thomas Commerford Martin, *Edison, His Life and Inventions*, 45–46. Some biographers think the book was actually written by William H. Meadowcroft, Edison's secretary. Edison told this particular story many times.
2. Ibid, 46.
3. John Dos Passos, quoted in Wyn Wachhorst., *Thomas Alva Edison, an American Myth*, 81.
4. Wyn Wachhorst, *Thomas Alva Edison*, 25.
5. Matthew Josephson, *Edison: A Biography*, 153–54.

CHAPTER SEVEN

1. Fred C. Kelly, "Charles F. Brush, my Friend, the Man, and the Inventor," 3.
2. Robert Friedel and Paul Israel, *Edison's Electric Light*, 13.
3. Ibid., 94.
4. Ibid., 37–38.
5. Ibid., 100.
6. Ibid., 106.
7. Ibid., 94.
8. Ibid., 96.
9. Ibid., 100.
10. Ibid., 102.
11. Ibid., 102.

CHAPTER EIGHT

1. Marc J. Seifer, *Wizard: The Life and Times of Nikola Tesla: Biography of a Genius*, 16.
2. Josephson, 346.
3. Ibid., 346.
4. Henry Adams, *The Education of Henry Adams: An Autobiography*, 356.

CHAPTER NINE

1. Douglas Coe, *Marconi, Pioneer of Radio*, 84.
2. Seifer, 305.

CHAPTER TEN

1. Helen N. Fessenden, *Fessenden, Builder of Tomorrows*, 32.

BIBLIOGRAPHY

BOOKS

Adams, Henry. *The Education of Henry Adams: An Autobiography*. 1907. Boston: Houghton Mifflin, 1918.

Aitken, Hugh G. J. *The Continuous Wave: Technology and American Radio, 1900–1932*. Princeton, N.J.: Princeton University Press, 1985.

Baldwin, Neil. *Edison: Inventing the Century*. New York: Hyperion, 1995.

Blake, G. G. *History of Radio Telegraphy and Telephony*. London: Radio Press, 1926.

Bowen, Catherine Drinker. *The Most Dangerous Man in America: Scenes from the Life of Benjamin Franklin*. Boston: Little, Brown, 1974.

Brands, H. W. *The First American: The Life and Times of Benjamin Franklin*. New York: Doubleday, 2000.

Briggs, Charles F., and Augustus Maverick. *The Story of the Telegraph, and a History of the Great Atlantic Cable*. New York: Rudd & Carleton, 1858.

Bright, Charles. *The Story of the Atlantic Cable*. New York: D. Appleton, 1903.

Carosso, Vincent P., with Rose C. Carosso. *The Morgans: Private International Bankers, 1854–1913*. Cambridge: Harvard University Press, 1987.

Carter, Samuel. *Cyrus Field: Man of Two Worlds*. New York: G. P. Putnam's Sons, 1968.

Cavanaugh, Cam, Barbara Hoskins, and Frances D. Pingeon. *At Speedwell in the Nineteenth Century*. Rev. ed. Speedwell, N.J.: Historic Speedwell, 2001.

Cheney, Margaret. *Tesla: Man out of Time*. Englewood Cliffs, N.J.: Prentice-Hall, 1981.

Coe, Douglas. *Marconi, Pioneer of Radio*. New York: Julian Messner, 1943.

Colladay, Morrison, and John J. O'Neill. *Nikola Tesla: Incredible Scientist and Prodigal Genius: The Life of Nikola Tesla*. Kila, Mont.: Kessinger, n.d.

Conot, Robert. *A Streak of Luck*. New York: Seaview, 1979.

Crane, Frank. *George Westinghouse: His Life and Achievements*. New York: William H. Wise, 1925.

Davidson, George E. *Beehives of Invention: Edison and His Laboratories*. Washington D.C.: National Park Service, 1973.

Dibner, Bern. *The Atlantic Cable*. Norwalk, Conn.: Burndy Library, 1959.

Dunlap, Orrin E., Jr. *Marconi: The Man and His Wireless*. New York: Macmillan, 1937.

Dyer, Frank Lewis, and Thomas Commerford Martin. *Edison: His Life and Inventions*. New York: Harper & Brothers, 1910. Online at etext.lib.virginia.edu/toc/modeng/public/Dye1Edi.html and . . . Dye2Edi.html.

Fahie, J. J. *A History of Wireless Telegraphy, 1838–1899*. 1899. New York: Arno Press, 1971.

Fessenden, Helen M. *Fessenden, Builder of Tomorrows*. New York: Coward-McCann, 1940.

Field, Henry M. *History of the Atlantic Telegraph*. 2d ed. New York: Charles Scribner, 1867.

Field, Rufus Chester. *The Persistent Fool: The Life and Trials of Cyrus West Field, Layer of the Atlantic Cable*. Privately printed, 1962.

Friedel, Robert, and Paul Israel. *Edison's Electric Light: Biography of an Invention*. New Brunswick, N.J.: Rutgers University Press, 1986.

Garbedian, H. Gordon. *George Westinghouse: Fabulous Inventor*. New York: Dodd, Mead, 1943.

Hammond, John Winthrop. *Men and Volts: The Story of General Electric*. Philadelphia: J. B. Lippincott, 1941.

Hawks, Ellison. *The Romance and Reality of Radio*. London: T. C. and E. C. Jack, 1923.

Houston, Edwin J., and A. E. Kennelly. *Electric Arc Lighting*. 3d ed. New York: McGraw, 1906.

Hubert, Philip G., Jr. *Inventors*. Men of Achievement series. New York: Charles Scribner's Sons, 1896.

Hughes, Thomas P. *Networks of Power: Electrification in Western Society, 1880–1930*. Baltimore: Johns Hopkins University Press, 1983.

Jacot, B. L., and D. M. B. Collier. *Marconi — Master of Space: An Authorized Biography*. London: Hutchinson, 1935.

Jolly, W. P. *Marconi*. London: Constable, 1972.

Josephson, Matthew. *Edison: A Biography*. New York: McGraw-Hill, 1959.

Labaree, Leonard W., ed. *The Papers of Benjamin Franklin*. New Haven: Yale University Press, vols. 3 and 4, 1961.

Leupp, Francis E. *George Westinghouse: His Life and Achievements*. Boston: Little, Brown, 1918.

Lewis, Tom. *Empire of the Air: The Men Who Made Radio*. New York: Edward Burlingame, 1991.

Mabee, Carleton. *The American Leonardo: A Life of Samuel F. B. Morse*. New York: Alfred A. Knopf, 1943.

McDonald, Philip B. *A Saga of the Seas: The Story of Cyrus W. Field and the Laying of the First Atlantic Cable*. New York: Wilson-Erickson, 1937.

Metzger, Thomas. *Blood and Volts: Edison, Tesla, and the Electric Chair*. Brooklyn, N.Y.: Autonomedia, 1996.

Munro, John. *Heroes of the Telegraph*. London: Religious Tract Society, 1891.

Nikola Tesla Museum. *Tribute to Nikola Tesla*. Belgrade, 1961.

Nye, David E. *The Invented Self: An Anti-biography, from Documents of Thomas A. Edison*. Odense, Denmark: Odense University Press, 1983.

Oslin, George P. *The Story of Telecommunications*. Macon, Ga.: Mercer University Press, 1992.

Prout, Henry G. *A Life of George Westinghouse*. New York: Charles Scribner's Sons, 1922.

Rasch, Ewald. *Electric Arc Phenomena*. Translated by K. Tornberg. New York: D. Van Nostrand, 1913.

Reingold, Nathan, ed. *The Papers of Joseph Henry*. Washington D.C.: Smithsonian Institution Press, vol. 8, 1972.

Roller, Duane H. D. *The De Magnete of William Gilbert*. Amsterdam: Menno Hertzberger, 1959.

Runes, Dagobert D., ed. *The Diary and Sundry Observations of Thomas Alva Edison*. New York: Philosophical Library, 1948.

Russell, W. H. *The Atlantic Telegraph*. London: Day & Son, n.d.

Scroggie, M. G. *Foundations of Radio*. 6th ed. London: Iliffe; New York: Philosophical Library, 1957.

Segrè, Emilio. *From Falling Bodies to Radio Waves: Classical Physicists and Their Discoveries*. New York: W. H. Freeman, 1984.

Seifer, Marc J. *Wizard: The Life and Times of Nikola Tesla: Biography of a Genius*. Secaucus, N.J.: Birch Lake Press, 1996.

Shiers, George, ed. *The Development of Wireless to 1920*. New York: Arno Press, 1977.

Standage, Tom. *The Victorian Internet: The Remarkable Story of the Telegraph and the Nineteenth Century's On-Line Pioneers*. New York: Walker, 1998.

Tate, Alfred O. *Edison's Open Door: The Life Story of Thomas A. Edison, a Great Individualist*. New York: E. P. Dutton, 1938.

Tesla, Nikola. *Lectures, Patents, Articles*. Edited by Vojin Popović. Belgrade, 1956.

———. *My Inventions*. Zagreb: Skolska Knjiga, 1977.

Vail, Alfred. *The American Electro Magnetic Telegraph*. Philadelphia: Lea & Blanchard, 1845.

Van Doren, Carl. *Benjamin Franklin*. New York: Viking, 1933.

Wachhorst, Wyn. *Thomas Alva Edison, an American Myth*. Cambridge: MIT Press, 1981.

Woodbury, David O. *Elihu Thomson, Beloved Scientist, 1853–1937*. Boston: Museum of Science, 1960.

Zeidler, J., and J. Lustgarten. *Electric Arc Lamps: Their Principles, Construction, and Working*. New York: D. Van Nostrand, 1908.

ARTICLES

"A Look Back: How We Would Have Worked." *Lexington Herald-Leader*, December 3, 1999.

Manchester, Harland. "The Blacksmith of Brandon." *Reader's Digest*, December 1942.

Michalowicz, Joseph C. "Origin of the Electric Motor." *AIEE Journal*, November 1948.

Mullaly, John. "The First Atlantic Telegraph Cable." *Journal of the Franklin Institute* (Philadelphia), 1907.

"Recovery of the Lost Cable." *Monthly Science News* (London), n.d.

Rezneck, Samuel. "The First Electric Motor." *Rensselaer Review*, vol. 3, number 2, June 1966.

Robertson, Andrew W. "About George Westinghouse and the Polyphase Electric Current." *Newcomen Society* (Princeton, N.J.), 1943.

Schiffer, Michael Brian. "The Blacksmith's Motor." *Invention & Technology*, winter 1994.

Tesla, Nikola. "The Onward Movement of Man." *Century Illustrated Monthly*, June 1900.

Westerdahl, Carl A. "Catalyst for Progress: Stephen van Rensselaer III." *Rensselaer*, September 1998.

SPEECH

Martin, Thomas Commerford. "An Appreciation of Thomas Davenport." Delivered at Farmdale, Vt., September 28, 1910.

WEB PAGES

Akst, Daniel. "Exchanging the Past." From the *Standard*, August 9, 1999. www.thestandard.com/article/0,1902,5754,00.html.

Allen, Hector. "Move Over Marconi! The Story of Mahlon Loomis of Oppenheim, NY." www.rootsweb.com/~nyfulton/loomis.html.

"Batteries, Current, and Ohm's Law." physics.bu.edu/py106/notes/Ohm.html.

Beals, Gerald. "His Greatest Achievements." www.thomasedison.com/brockton.htm.

Belrose, John S. "Fessenden and Marconi: Their Differing Technologies and Transatlantic Experiments During the First Decade of this Century." ewh.ieee.org/reg/7/millennium/radio/radio_differences.html.

Benson, Theo. "Nikola Tesla." www.theelectricchair.com/tesla_article_by_theo_benson.htm.

Blume, Harvey. "'Autism & the Internet' or 'It's the Wiring, Stupid.'" www.mit.edu/comm-forum/papers/blume.html.

Brain, Marshall. "How Electric Motors Work." www.howstuffworks.com/motor2.htm.

Calvert, John. "Electrical Machinery." www.du.edu/~jcalvert/tech/elmotors.htm.
———. "The Electromagnetic Telegraph." www.du.edu/~jcalvert/tel/morse/morse.htm.

Cameron, Ewan, and Paul May. "Nitrous Oxide — Laughing Gas." mole.chm.bris.ac.uk/motm/n2o/n2oj.htm.

Carlson, W. Bernard. "Nikola Tesla, Illusion, and Invention in Nineteenth Century Electricity." cs.muohio.edu/~bhc/Conference/carlson.html.

Casson, Herbert N. "The Telephone and National Efficiency." Chapter 7 of *The History of the Telephone* (Chicago: A. C. McClurg, 1910). www.worldwideschool.org/library/books/tech/engineering/TheHistoryoftheTelephone/chap7.htm.

Catholic Encyclopedia. "Edouard Branly." www.newadvent.org/cathen/02740a.htm.

"Commutator and Brushes on DC Motor." www.hyperphysics.phy-astr.gsu.edu/hbase/magnetic/comtat.html.

Currier, Dean P. "A Biographical History of Induction Coils." www.radiantslab.com/quackmed/Deanbio.html.

Czitrom, Daniel J. Chapter 1 of *Media and the American Mind: From Morse to McLuhan* (Chapel Hill: University of North Carolina Press, 1982). www.uniworld.hu/egyetem/polfiz/irodalom/cikk25.htm.

"Did a West Virginian Invent Radio?" From the Beckley, W.Va., *Raleigh Register*, September 7, 1976. members.aol.com/jeff560/loomis.html.

Donway, Roger. "Benjamin Franklin: Enlightenment Archetype." From *Navigator* 3, no. 2 (2000). www.objectivistcenter.org/articles/donway_benjamin-franklin.asp.

"Dr. Charles Jackson: Gesner's Plagarism." www.saintjohn.nbcc.nb.ca/~Heritage/AlbertMines/Plagarism.htm.

"Dr. Nikola Tesla." www.frank.germano.com/nikolatesla.htm.

Durant, Will, and Ariel Durant. Condensed from *The Story of Civilization*, vol. 9, *The Age of Voltaire* (New York: Simon & Schuster, 1965), pp. 520–23. www.owecc.net/crunyon/CE/Frankenstein/Name/FranklinElect.html.

Encyclopedia Britannica. "Electromagnetism." www.britannica.com/bcom/eb/article/2/0,5716,108502+16,00.html.

"Expériences et Observations sur l'Electricité faites à Philadelphie en Amérique." www.netrax.net/~rarebook/s980318.htm.

"Extending Man's Voice by Wire and Radio." Based on *Electronic Design* 24, no. 4 (1976). www.luminet.net/~wenonah/history/edpart3.htm.

"Fessenden." www.radiocom.net/Fessenden.

"The Fight for AC." Chapter 2 of *Tesla: The Electric Magician.* www.parascope.com/en/0996/tesla2.htm.

First Spiritual Temple. "Sir Oliver Lodge." www.fst.org/lodge.htm.

"First Transatlantic Transmissions." www.marconicalling.com/museum/html/events/events-i=30-s=0.html.

Fist, Stewart. "Morse and Signal Regeneration: Long-Distance Legacy." From the *Australian*, November 2, 1999. www.electric-words.com/hist/991102history.html.

"The Forgotten Canadian." www.kwarc.on.ca/fessenden.html.

"The Forgotten Genius." Chapter 6 of "Tesla: The Electric Magician." www.parascope.com/en/0996/tesla6.htm.

Fowler, Michael. "Historical Beginnings of Theories of Electricity and Magnetism." galileoandeinstein.physics.virginia.edu/more_stuff/E&M_Hist.html.

Franklin, H. Bruce. "*Billy Budd* and Capital Punishment: A Tale of Three Centuries." From *American Literature*, June 1997. newark.rutgers.edu/~hbf/bbcap.htm.

"Free Energy." Chapter 3 of *Tesla: The Electric Magician.* www.parascope.com/en/0996/tesla3.htm.

Fry, Mervyn C. "Radio's First Voice . . . Canadian!" www.ewh.ieee.org/reg/7/millennium/radio/radio_birth.html.

"The Guglielmo Marconi Case: Who Is the True Inventor of Radio?" www.mercury.gr/tesla/marcen.html.

Herskovits, Zara. "To the Smithsonian or Bust: The Scientific Legacy of Nikola Tesla." From *Yale Scientific* 71 (1999). www.yale.edu/scimag/Archives/Vol71/Tesla.html.

"His Wildest Dreams." Chapter 5 of "Tesla: The Electric Magician." www.parascope. com/en/0996/tesla5.htm.

"A History of Anaesthesia at Harvard University." www.hmcnet.harvard.edu/ anesthesia/history/vandam.html.

"History of Magnetics." www.ocean.washington.edu/people/grads/mpruis/magnetics/ history/hist.html.

Hochfelder, David. "Joseph Henry: Inventor of the Telegraph?" www.si.edu/ archives/ihd/jhp/joseph20.htm.

"A Hole in the Floor." www.150.si.edu/chap2/two.htm.

"The Invention of the Electron Tube." www.netsonian.com/antiqueradio/ radiodocs/tubedev.htm.

Jehl, Francis. "Thomas Alva Edison in Menlo Park, NJ." Excerpts from *Menlo Park Reminiscences,* 3 vols. (Dearborn, Mich.: Edison Institute, 1937–41). www.jhalpin.com/metuchen/tae/jehl.htm.

Kelly, Fred C. "Charles F. Brush, my Friend, the Man, and the Inventor." www.ameritech.hel/jess_laFavre/brushbiz.htm.

Kryzhanovsky, Leonid, and James P. Rybak. "Recognizing Some of the Many Contributors to the Early Development of Wireless Telegraphy." www.ptti.ru/ eng/forum/article1.html.

Lienhard, John H. "Electric Lights Before Edison." Engines of Our Ingenuity, no. 1330. www.uh.edu/engines/epi1330.htm.

———. "Fleming's Electric Valve." Engines of Our Ingenuity, no. 1323. www.uh.edu/engines/epi1323.htm.

———. "Franklin's Electricity." Engines of Our Ingenuity, no. 510. www.uh.edu/ engines/epi510.htm.

———. "Indian Telegraph." Engines of Our Ingenuity, no. 54. www.uh.edu/ engines/epi54.htm.

———. "Inventing the Telegraph." Engines of Our Ingenuity, no. 1393. www.uh. edu/engines/epi1393.htm.

———. "Nikola Tesla." Engines of Our Ingenuity, no. 174. www.uh.edu/engines/ epi174.htm.

———. "Perry Collins." Engines of Our Ingenuity, no. 772. www.uh.edu/engines/ epi772.htm.

———. "William Gilbert." Engines of Our Ingenuity, no. 613. www.uh.edu/ engines/epi613.htm.

"Lightbulbs: Edison Didn't Invent Them, and What It Means to Be 'Westing-housed.'" home.nycap.rr.com/useless/lightbulbs/

Lomas, Robert. "The Lost Prophet of Electrical Science." www.lauralee.com/ news/lostprophet.htm.

McAuliffe, Kathleen. "The Undiscovered World of Thomas Edison." From the *Atlantic,* December 1995. www.theatlantic.com/issues/95dec/edison/edison.htm.

Macdonald, D. L., and Kathleen Scherf, eds. From the introduction to *Franken-stein*, by Mary Shelley (Peterborough, Ont.: Broadview, 1996). www.towson. edu/~sallen/311/Frankenstein.html.

McEwen, Neal. "Morse Code or Vail Code? Did Samuel F. B. Morse Invent the Code as We Know it Today?" www.metronet.com/~nmcewen/vail.html.

Mazlish, Bruce. "The Man-Machine and Artificial Intelligence." www.stanford. edu/group/SHR/4-2/text/mazlish.html.

Miall, David S. "Frankenstein Presentation." www.arts.ualberta.ca/~dmiall/ romant96/FRANKPRJ.HTM.

Middlekauff, Robert. Chapter 1 of *Benjamin Franklin and His Enemies* (Berkeley: University of California Press, 1996). www.washingtonpost.com/wpsrv/ style/longterm/books/chap1/benjaminfranklinandhis enemies.htm.

Mishkind, Barry. "This Is the Broadcast History Section of the Broadcast Archive." www.oldradio.com/current/bc_roots.htm.

Mollet, Dennis, and James Kistner. "Lightning." www.public.asu.edu/~gbadams/ lightning/lightning.html.

Morse, Robert Eaton. "Samuel Finley Breeze Morse: Artist-Inventor." www.redlands fortnightly.org/morse.htm.

"Morse Code or Vail Code? Did Samuel F. B. Morse invent the code as we know it today? Franklin Pope and William Baxter give some answers." www.metronet. com/~nmcewen/vail.html

"Nikola Tesla: Chicago World's Fair." www.neuronet.pitt.edu/~bogdan/tesla/ chicago.htm.

"19th Century Physical Sciences 1 — Romantic Forces in Chemistry and Elec-tromagnetism: How Davy and Oersted vanquished imponderable fluids and replaced them with interconvertible forces." www.students.ou.edu/K/ Gary. M.Kroll-1/lecture_5a.html.

"Ohm's Law." www.a-s.clayton.edu/pratte/jmp8.html.

"Ohm's Law." www.the12volt.com/ohm/ohmslaw.asp.

"One Story of Nikola Tesla." www.flyingmoose.org/truthfic/tesla.htm.

"Origin of Electric Power." www.americanhistory.si.edu/csr/powering/prehist/ prehist.htm.

"Oxygen Therapy: The First 150 Years (Continued)." www.mtsinai.org/ pulmonary/papers/ox-hist/ox-hist1.html.

Pedersen, Fran. "The Alaska-Siberia." *Alaska Science Forum*, article 68, October 25, 1977. www.gi.alaska.edu/ScienceForum/ASF0/068.html.

"Phase and Polarity." www.vansevers.com/Notes/Phase/phase.html.

Phelan, Thomas, and D. Michael Ross. "Amos Eaton and the Magnificent Ex-periment." From *Rensselaer*, December 1998. www.rpi.edu/dept/NewsComm/ Magazine/Dec98/eaton2.html.

"PHYS1259 Lecture 3 (Magnetism)." www.phys.unsw.edu.au/~epe/1259.L3/1259.L3.html.

Pierpont, William G. From "A Brief History of Morse Telegraphy," chapter 19 of The Art and Skill of Radio-Telegraphy. www.geocities.com/gm0rse/n0hff/c19a.htm.

Pope, Frank L. Chapter 3 of Modern Practice of the Electric Telegraph, 11th ed. (New York: D. Van Nostrand, 1881). www.la.znet.com/~cdk14568/mpet/chap3.html.

"Radio's Version of 'Who's on First?' Many Claims Have Been Made, but Radio's Paternity Is Still a Question." www.nrcdxas.org/articles/who1st.txt.

"Re: Henwood on Keynes." www.csf.colorado.edu/forums/pkt/oct98/0215.html.

Rybak, James P. "Alexander Popov: Russia's Radio Pioneer." www.ptti.ru/eng/forum/article2.html.

Saunders, Melvin D. "Wireless Electricity of Nikola Tesla." www.braincourse.com/wirelessa.html.

Scott, Carole E. "The Radio Inventor/Entrepreneurs." www.westga.edu/~bquest/2001/radio.htm.

Seckel, Al, and John Edwards. "Franklin's Unholy Lightning Rod." www.evolvefish.com/freewrite/franklgt.htm.

Seitz, Frederick. "The Project: A Physicist's Perspective." www.si.edu/archives/ihd/jhp/projec02.htm.

Sherman, Roger. "Joseph Henry's Contributions to the Electromagnet and the Electric Motor." www.si.edu/archives/ihd/jhp/joseph21.htm.

"A Short History of Electric Machines." www.historia.et.tudelft.nl/pub/art/machines.php3.

Shulman, Seth. "Unlocking the Legacies of the Edison Archives." www.technology review.com/articles/shulman0297.html.

"Sir Oliver Lodge." www.netcentral.co.uk/steveb/focus/003.htm.

Smith, Brian. "The Story of Reginald Aubrey Fessenden." www181.pair.com/otsw/Fessenden.html.

Stern, David P., and Mauricio Peredo. "Dr. William Gilbert's 'Terrella.'" www.acmi.net.au/AIC/TERRELLA.html.

"The Story of the Telegraph." www.acmi.net.au/ACI/TELEGRAPHY_LULA.html.

"Stubblefield's Wireless." www.anomalyinfo.com/sa00005b.shtml.

Symons, E. D. P. "A Magnificent Failure — Edward Davy (1806–1885)." From IEE Review (Institution of Electrical Engineers, London), May 1996. www.asap.unimelb.edu.au/bsparcs/other/iee_davy.htm.

"Tesla, the Greatest Hacker of All Time." Chapter 9 of "Tesla: The Electric Magician." www.parascope.com/en/0996/tesla9.htm.

"Tesla's Death Ray." Chapter 4 of "Tesla: The Electric Magician." www.paras-cope. com/en/0996/tesla4.htm.

"Timeline: Ezra Cornell: A Nineteenth Century Life." www.rmc.library.cornell. edu/Ezra-exhibit/time/ECfa.timeline.html.

"TWP Question and Answer Forum." www.pages.nyu.edu/~jas4/questions.htm.

Visser, Thomas D. "Smalley-Davenport Shop, Forestdale, Vermont." www.uvm. edu/~histpres/SD/hist.html.

"What is Inductance?" In "Battery Cable Inductance." www.traceengineering. com/technical/tech_notes/tn8.html.

Wicks, Frank. "The Blacksmith's Motor." From *Mechanical Engineering*, July 1999. www.memagazine.org/backissues/july99/features/blacksmith/blacksmith.html.

———. "Full Circuit." From *Mechanical Engineering*, September 2000. www.memagazine.org/backissues/sept00/features/full/full.html.

"William Thomas Green Morton (1819–1868): American dentist, who claimed to be the discoverer of ether." www.aana.com/archives/imagine/ 1996/10imagine96.asp

Wunsch, A. David. "Misreading the Supreme Court: A Puzzling Chapter in the History of Radio." www.mercurians.org/nov98/misreading.html.

INDEX